Smart Innovation, Systems and Technologies

Volume 416

The Smart Innovation, Systems and Technologies book series encompasses the topics of knowledge, intelligence, innovation and sustainability. The aim of the series is to make available a platform for the publication of books on all aspects of single and multi-disciplinary research on these themes in order to make the latest results available in a readily-accessible form. Volumes on interdisciplinary research combining two or more of these areas is particularly sought.

The series covers systems and paradigms that employ knowledge and intelligence in a broad sense. Its scope is systems having embedded knowledge and intelligence, which may be applied to the solution of world problems in industry, the environment and the community. It also focusses on the knowledge-transfer methodologies and innovation strategies employed to make this happen effectively. The combination of intelligent systems tools and a broad range of applications introduces a need for a synergy of disciplines from science, technology, business and the humanities. The series will include conference proceedings, edited collections, monographs, handbooks, reference books, and other relevant types of book in areas of science and technology where smart systems and technologies can offer innovative solutions.

High quality content is an essential feature for all book proposals accepted for the series. It is expected that editors of all accepted volumes will ensure that contributions are subjected to an appropriate level of reviewing process and adhere to KES quality principles.

Indexed by SCOPUS, EI Compendex, INSPEC, WTI Frankfurt eG, zbMATH, Japanese Science and Technology Agency (JST), SCImago, DBLP.

All books published in the series are submitted for consideration in Web of Science.

Shih-Pang Tseng • Anand Paul • Jeng-Shyang Pan • Margarita Favorskaya
Editors

Advances in Intelligent Information Hiding and Multimedia Signal Processing, Volume 2

Proceeding of the 19th International Conference on IIH-MSP in conjunction with 11th International Conference on Orange Technology, Applications and Tools, December 5-7, 2023, Daegu, South Korea

Springer

Editors
Shih-Pang Tseng
Changzhou College of Information Technology
Changzhou, China

Jeng-Shyang Pan
College of Computer Science and Engineering
Shandong University of Science and Technology
Shandong, China

Anand Paul
Kyungpook National University
The School of Computer Science and Engineering
Daegu, Korea (Republic of)

Margarita Favorskaya
Reshetnev Siberian State University of Science and Technology
Krasnoyarsk, Russia

International Conference on Intelligent Information Hiding and Multimedia Signal Processing iihmsp iihmsp2023 IIHMSP 19 Daegu, Korea (Republic of) 2023 12 4 2023 12 7 https://icot-2023.com/

ISSN 2190-3018 ISSN 2190-3026 (electronic)
Smart Innovation, Systems and Technologies
ISBN 978-981-97-8759-3 ISBN 978-981-97-8760-9 (eBook)
https://doi.org/10.1007/978-981-97-8760-9

This Springer imprint is published by the registered company Springer Nature Singapore Pte Ltd.
The registered company address is: 152 Beach Road, #21-01/04 Gateway East, Singapore 189721, Singapore

Preface

This volume composes the proceedings of the 19th International Conference on Intelligent Information Hiding and Multimedia Signal Processing (IIHMSP-2023) and 11th International Conference on Orange Technology (ICOT-2023) hosted by Kyungpook National University and held in Daegu, South Korea, on December 4–7, 2023. ICOT-2023 and IIHMSP-2023 is technically co-sponsored by Springer, Kyungpook National University, Tainan Information Software Association, Shandong University of Science and Technology, and National Cheng Kung University. The topics cover classical subjects such as information hiding and security, multimedia signal processing and networking, computing for healthcare, affective computing, and bio-inspired multimedia technologies and systems. The conference also contains quite a number of papers dealing with broad aspects of artificial intelligence.

Fifty-eight papers were accepted in these proceedings. We would like to thank the authors for their tremendous contributions. We would also express our sincere appreciation to the reviewers, the program committee members, and the local committee members for making this conference successful. Finally, we would like to express our special thanks for the great help from Kyungpook National University for locally organizing the conference.

Changzhou, China
Daegu, South Korea
Qingdao, China
Krasnoyarsk, Russia
December 2023

Shih-Pang Tseng
Anand Paul
Jeng-Shyang Pan
Margarita Favorskaya

Contents

Chapter 1
An Empirical Study of Human-Machine Collaboration Teaching on English Listening Anxiety among Non-English Majors

Jiamei Fu, Hongfang Yan, and Jun Zheng

Abstract Listening anxiety is believed to affect English learning behavior and outcomes of non-English majors. This paper reports an empirical study of applying human-machine collaboration in the teaching of college English listening. A framework of human-machine collaboration in classroom teaching is provided, and the questionnaire and SPSS software are adopted to analyze the data. The analysis results show that human-machine collaboration teaching is effective and efficient to alleviate non-English majors' listening anxiety from the dimensions of lack of listening skills, listening result worry, and listening process fear. Low achievers worry too much over the listening results, and the difference in test scores mainly results from lack of listening skills.

1.1 Research Background

Anxiety is not only an important affective factor in foreign language learning, but also one of the main triggering factors affecting foreign language learners' learning behavior and outcomes [1]. It was found that for non-English major college students, among five dimensions of the foreign language anxiety construct, the level of classroom anxiety ranked first, followed by speaking anxiety, listening anxiety, writing anxiety, and reading anxiety [2]. And how to alleviate listening anxiety is a recognized challenge. *The Basic Information of National Education Development in 2022* released by the Ministry of Education of the People's Republic of China in 2023 shows that the average student-teacher ratio in higher education in China

---

This paper is the periodical achievement of School Research Fund (2022YB23).

J. Fu · H. Yan · J. Zheng (✉)
School of General Education, Sanda University, Shanghai, China
e-mail: jmfu@sandau.edu.cn; zhengj@sandau.edu.cn

S.-P. Tseng et al. (eds.), *Advances in Intelligent Information Hiding and Multimedia Signal Processing, Volume 2*, Smart Innovation, Systems and Technologies 416, https://doi.org/10.1007/978-981-97-8760-9_1

is 23.5:1. Teachers are unable to pay attention to students' individual differences and address their listening anxiety due to limited time and energy. With the rapid development of information technology and its application in foreign language teaching, foreign language teaching has undergone a revolution, and has made significant innovations in terms of teaching philosophy, methods, model, system of evaluation, and management [3, 4]. Human-machine collaboration (HMC) teaching helps to realize large-scale personalized and precise education, and can maximize the advantages of artificial intelligence in education [5, 6].

Therefore, this paper tries to explore the following three research questions:

- What are the features of non-English majors' listening anxiety?
- How to apply human-machine collaboration during classroom teaching?
- Are there positive effects of human-machine collaboration teaching on students' English listening anxiety? What are they?

1.2 Human-Machine Collaboration and Its Application

1.2.1 Human-Machine Collaboration (HMC)

Educational artificial intelligence refers to a new research and application paradigm which applies artificial intelligence and big data technology in the field of education featuring the integration of technology into the classroom.

One typical application of educational artificial intelligence which features the integration of technology into the classroom is human-machine collaboration. Qian et al. (1990) [7] defined human-machine collaboration as a collaborative effort between human and machine, each exerting its own advantages. Human-machine collaboration teaching is the interaction between human teachers and machines to accomplish instructional goals by learning from each other, jointly to improve the quality of teaching and learning [8].

Figure 1.1 shows the typical framework of human-machine collaboration teaching in which the teacher is the guide, students the center, and AI enables smart teaching and learning in and out of classroom so as to meet students' precise needs.

1.2.2 Application of HMC in English Listening Teaching

Several teachers have already integrated one platform or app into the human-machine collaboration in English context, such as iWrite. [9, 10] In this experimental school year, the available smart platforms and apps that both teachers and students can easily access are Treenity, Renren Dictionary, Snap English, and Quizlet. The detailed teaching design and application are shown in Table 1.1. By playing to their strengths, teachers and AI cooperate to set listening objectives based

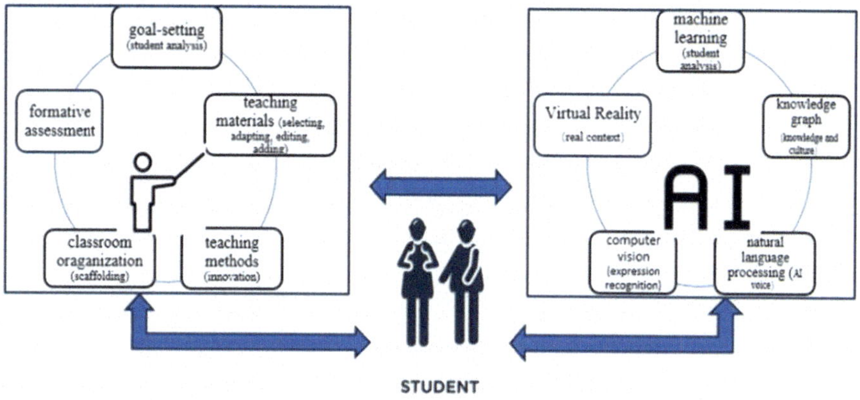

Fig. 1.1 Typical framework of HMC teaching

on the anxiety scale and promote English listening teaching and learning in and out of classroom effectively and efficiently.

To get students prepared, at the very beginning, the teacher will show students how to use the platforms and apps. In the following studies, students will be proficient in using them and get used to different modes like text, audio, and video which have been proved to be of great help for listening improvement [11].

1.3 Research Design and Data Collection

1.3.1 Subjects

There are 130 valid subjects (28 males and 102 females) from one university in Shanghai. They are non-English major freshmen and have had listening instruction in their integrated English class for two terms. All the subjects attend the class 4 periods (180 min) per week within 32 weeks, which ensures human-machine collaboration teaching every week and that valid data could be obtained for this study.

1.3.2 Research Design

Researchers from home and abroad construct listening anxiety assessment dimensions from different perspectives. Referring to the Foreign Language Listening Anxiety Scale (FLLAS) designed by Elkhafaifi [12–14] and its analyses by Xian Zhang and Yu Shi et al. [15, 16], the writers translated the FLLAS from English into Chinese so that all the subjects could fully and correctly understand all the questions.

Table 1.1 The application of HMC in English listening teaching

Period	Objectives	Content	What teacher does	What AI does
Before class	Detect and set target vocabulary and skills	1. Vocab. Pre-test 2. CET 4 listening model test	1. Manage the class online 2. Set listening objectives 3. Make teaching plan	Treenity: 1. Test bank 2. Test release, grading and result analysis
During class	Know about different topics and cultural background	Related topics and culture in textbook, CET 4 and China daily	1. Select, revise and add the materials 2. Manage resources, tests and knowledge on Treenity	Treenity: All the audios, videos, texts and tests, and their relations
	Listen extensively for different pronunciations/tones/voices, topics and main ideas	Textbook, CET 4 and China daily	1. Edit and categorize materials in the form of text 2. Operate snap English to produce proper AI voice	Snap English: Produce AI voices like American/British woman and man, American/British boy and girl, American narrative/calm/fast/soft
	Listen intensively for target vocabulary in terms of single word, chunk, sentences, and passage, the logic flow in context and useful detailed information	The above materials are shown repeatedly in different ways: 1. Word (symbolic associations) 2. Chunk (syntagmatic and paradigmatic associations) 3. Sentence (pragmatic associations)	1. Manage multi-modal tasks in different forms and difficulty levels (sensory, motor activities, and cognitive activities) 2. Scaffold: Observe emotions and performance, identify the difficulty, play the audio more times, pause where necessary, etc.	1. Snap English recognizes pictures and texts to produce AI voice. 2. Treenity grades the tests in class Immediately.

	Get used to and identify listening skills such as elision, weak pronunciation, and swallowing sound	The same materials as above	Add comma to long sentences on snap English to help AI make pauses	The same as above
	Be used to speaking speed, esp. in CET 4 listening section	The same materials as above	1. Observe students 2. Adjust the speaking speed when playing the audio	Snap English produces slow, medium, normal, fluent, and fast speaking speed.
	Lessen mother tongue interference in the real context	Clip video of American drama with related topic, vocabulary, and culture	1. Select, revise, and add teaching materials 2. Guide students to the flow of ideas	Renren dictionary provides mass authentic videos and bilingual subtitles.
After class	Review target vocabulary (long-term memory and immediate response)	1. Target vocabulary. 2. Listening materials.	1. Manage class online 2. Record the performance of students on and offline	1. Quizlet provides quick, repetitive, and mass gamified exercises, such as flashcards and bingo. 2. Knowledge graph on Treenity

Based on it, this paper will explore the following five dimensions according to the actual situation: listening process fear, listening result worry, lack of listening confidence, lack of listening skills, and listening cognitive anxiety.

- Listening process fear refers to students' anxious reactions to difficulties in word sound-meaning and peer pressure during the listening.
- Listening result worry refers to students' anxious feelings due to negative attitude and hurt of poor comprehension.
- Lack of listening confidence describes students' lack of confidence and dissatisfaction with their listening due to bad habits and insufficient motivation.
- Lack of listening skills refers to students' failure to acquire listening skills through previous knowledge study and training, such as identifying different language units, logical training, cultural accumulation, and countering effect of mother tongue.
- Listening cognitive anxiety refers to anxiety about the cognitive construct component (including the cognitive component of foreign language skills), as foreign language learners with lower competency tend to be more anxious about the difficulty scales and learning new knowledge.

Questionnaires were distributed to students regarding the five influencing factors and the two-level indicators, and the correspondence table with the questionnaire items is shown in Table 1.2.

Table 1.2 Correspondence table between three-level indicators of variables and question items

One-level indicators	Two-level indicators	Question items	Indicator attribute
Listening process fear	Fear about new word sound-meaning Peer pressure	7,8,10 17	Negative, quantitative
Listening result worry	Negative attitude Poor comprehension	4 1,5,6	Negative, quantitative
Lack of listening confidence	Bad habits Insufficient motivation	14 12,13,18	Negative, quantitative
Lack of listening skills	Poor logic Lack of cultural background Negative effect of language unit and mother tongue	2,3 19 9	Negative, quantitative
Listening cognitive anxiety	Anxiety about learning new knowledge Anxiety about difficulty Scales	11,20 15,16	Negative, quantitative

1.3.3 Data Collection

The study lasted for two semesters, and the questionnaires were distributed to the subjects twice, namely in the week of pre- and post-course respectively. The subjects majored in art (23), education (22), management (37), economics (43), science and technology (1), and language and literature (4). A total of 130 valid questionnaires were distributed and collected. The subjects had one listening pre-tests at the start of the course, four during the term, and one in the final. All the tests shared the same question types and the levels of difficulty were graded to match students' listening competency at each period.

1.4 Research Result and Analysis

This questionnaire measured the impact of human-machine collaboration teaching on listening anxiety based on five factors extracted from the questionnaire items. SPSS software was used to conduct a factor analysis which is reliable and valid (Cronbach's a of most factors >0.7, the cumulative variance contribution rate > 66%).

1.4.1 General Analysis

The five influencing factors were scored on a scale of 0 to 20, with higher values indicating a greater degree of negativity. As shown in Fig. 1.2, a comparative analysis of the median scores of the five influencing factors revealed that lack of listening skills decreased from 15 to 11, listening result worry decreased from 15 to 13, and listening process fear decreased from 14 to 13, while lack of listening confidence and listening cognitive anxiety did not change much.

As shown in Table 1.3, the average score of listening pre-test is 57.738 with a standard deviation of 21.2269, and the average score of listening post-test is 71.323 with a standard deviation of 12.2279. The listening post-test scores covered four test scores (accounting for 40%) during the term and the final score (accounting for 60%). Students' listening scores have greatly improved, and class scores are more stable.

1.4.2 Correlation Analysis

The correlation analysis revealed a significantly positive relationship between listening process fear, listening result worry, lack of listening confidence, and listening

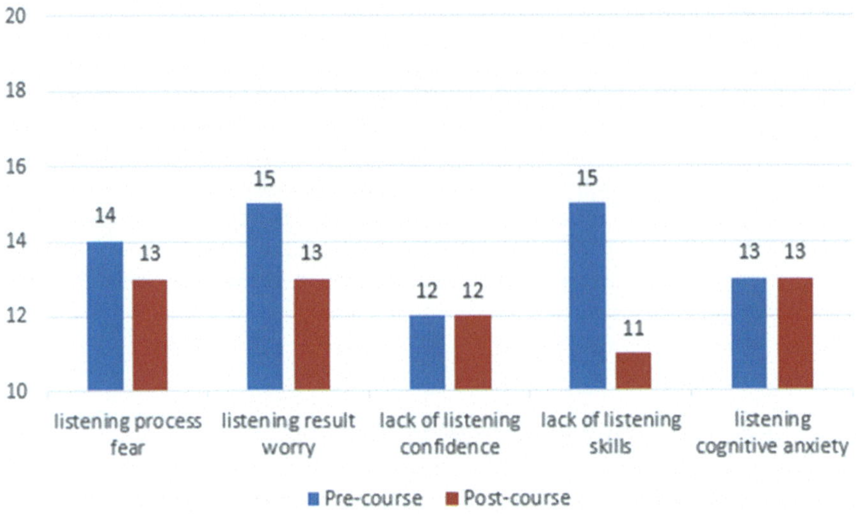

Fig. 1.2 Histogram of one-level indicators comparison between pre and post course

Table 1.3 Score analysis

		Listening pre-test	Listening post-test
N	Valid	130	130
	Missing	0	0
Mean		57.738	71.323
Median		60.000	73.000
Std. deviation		21.2269	12.2279
Variance		450.582	149.523
Skewness		−0.134	−0.542
Std. errors of skewness		0.212	0.212
Kurtosis		−0.739	0.552
Std. errors of kurtosis		0.422	0.422

cognitive anxiety (correlation coefficients > 0.6). The factors were negatively correlated with scores with a significantly negative relationship between lack of listening skills and scores (correlation coefficient = −0.715).

1.4.3 K-means Clustering Analysis

K-means clustering was conducted based on the test score and overall factor score which was calculated after analyzing 20 two-level indicators. Three clusters were obtained as shown in Table 1.4. The data distribution is shown in Fig. 1.3.

Table 1.4 Final cluster center

	Cluster		
	1	2	3
Test scores	81	47	64
Overall factor score	−0.09	0.18	0.08

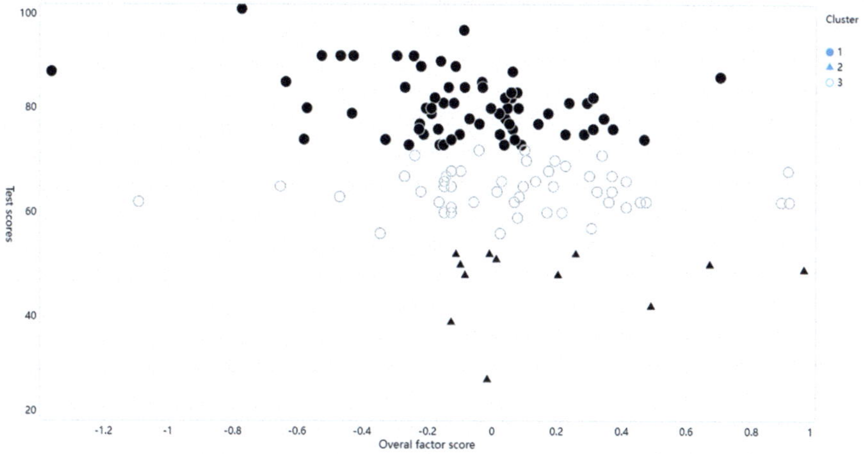

Fig. 1.3 Scatter plot

The average values of each two-level indicator for three different student groups are shown in Fig. 1.4. There are 20 questions in the questionnaire, and the average performance of students in three clusters (in blue, red, and green respectively) shows the differences and reasons.

1.5 Conclusion and Discussion

This study explores the influence of human-machine collaboration teaching on English listening anxiety among non-English major college students based on the features of their listening anxiety. The results show that:

- Human-machine collaboration teaching can moderate students' listening anxiety to great extent, and students have made significant progress in listening tests. In terms of these positive changes, the factor of lack of listening skills ranks the first. Students could recognize different units of language, capture the key information with better logic (such as the ability to predict), and establish a better intercultural awareness. Listening result worry and listening process fear both rank the second. Although students practice their listening all together in the same class, they learn to avoid being distracted and concerning too much over the score. They could also consolidate their knowledge about words, which

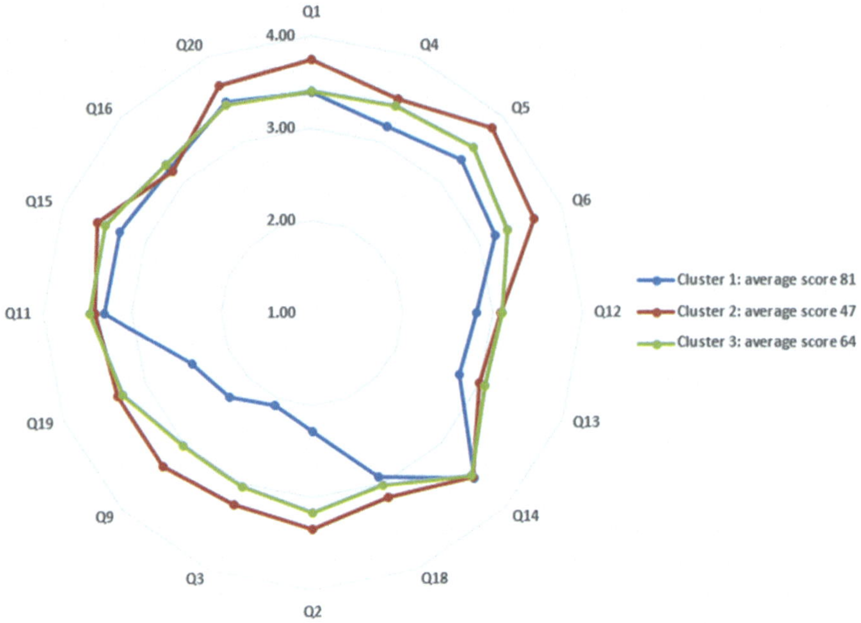

Fig. 1.4 Radar plot

proves that human-machine collaboration teaching can help students learn the pronunciation, semantic meanings, and part of speeches of a word.

- It is revealed by comparing listening anxiety characteristics of the three student groups (students who are good, passed, and failed respectively) that students who failed listening tests had a serious psychological burden and needed to adjust their mindset and build up their psychology before listening. The reason for the obvious difference in test scores was the lack of listening skills, while the listening cognitive anxiety and the lack of listening confidence have a significantly negative impact on the test scores.
- The lack of listening confidence and listening cognitive anxiety, which are regarded as value factors such as habits, motivation, and cognition, did not change much compared to the knowledge and skills that can be accumulated over a certain period of time, which can be interpreted as human-machine collaboration teaching moderating listening anxiety from the outside in.

It is recommended that human-machine collaboration teaching strengthen the interference on the lack of listening confidence and listening cognitive anxiety. For students' listening confidence building, human-machine collaboration teaching should help them form good learning habits, conform to students' cognitive development laws, and stimulate their motivation step by step. For listening cognitive anxiety relief, human-machine collaboration teaching should make full use of multimodal teaching mode to help students step out of their comfort zone by transiting from single text mode to bi-modal mode including both text and audio, then to single audio mode. In this way, students can gradually overcome their cognitive anxiety.

There is still room to improve. For example, human-machine collaboration can be applied further on how to monitor all the students' learning process effectively and efficiently.

References

1. Liru, B.: On the inducing factors and performance representations of English writing anxieties from college English learners [J]. Foreign Lang. Lit. **37**(5), 129–138 (2021). https://doi.org/10.3969/j.issn.1674-6414.2021.05.017
2. Yan, G., Jinfen, X.: A multi-dimensional study on foreign language anxiety of non- English major college students in China [J]. Foreign Lang. World. **163**(04), 2–11 (2014)
3. Liying, C.: A study on blended college English teaching under information technology environment—comment on *a study of blended learning in college English teaching* [J]. High. Educ. Explor. **1**, 133 (2020). https://doi.org/10.3969/j.issn.1673-9760.2020.01.023
4. Linsen, L., Jie, Y.: Evaluation and analysis of college English curriculum from the perspective of learning experience [J]. J. Southwest Univ. (Soc. Sci. Ed.). **49**(2), 202–214 (2023). https://doi.org/10.13718/j.cnki.xdsk.2023.02.018
5. Zhou Yueliang, Wu Yinhe, Cai Lianyu. Research on Teacher Education Reform Oriented to Human-Computer Collaborative Education[J]. e-Education Res., 2022,43(10):5–11
6. Kaiquan, C., Xiaoli, H., Zhanfei, Z., Liu Xingli, H., Xiaosong.: The analysis of construction and application mode of scenario construction in intelligent education from the perspective of human-machine cooperation—a review of domestic and international review on human-machine cooperative education in the past ten years [J]. J. Dist. Educ. **40**(02), 3–14 (2022)
7. Qian Xuesen, Y., Jingyuan, D.R.: A new scientific area— open and complex system and its methodology [J].Chinese journal of. Nature. **01**, 3–10+64 (1990)
8. Xuejiao, F., Mingxing, Z., Youfu, Z.: Design on the overall framework and key links of human-computer collaborative precision teaching [J]. Open Educ. Res. **29**(02), 91–102 (2023)
9. Hua-li, Y., Ying, G., Tao, H., Yuan, Z.: Case study of primary school English writing teaching with the support of human-machine collaboration [J]. Mod. Educ. Technol. **30**(04), 74–80 (2020)
10. Yingliang, L., Shenglan, L., Jincai, Y.: Man-machine cooperative teaching and its application from the sociaocultural activity-theory perspective: a case of iWrite-assisted English writing instruction [J]. China Educ. Technol. **430**(11), 108–116 (2022)
11. Miao, T., Meng, W., Zi-hang, Z.: An empirical study of the effects of multi-mode flipped classroom teaching mode on college students' English listening comprehension [J]. Heilongjiang Res. High. Educ. **37**(11), 152–156 (2019)
12. Bandura, A.: Self-efficacy: towards a unifying theory of behavior change [J]. Psychol. Rev. **2**, 191–215 (1977)
13. MacIntyre, P.D.: How does anxiety affect second language learning? A reply to Sparks and Ganschow [J]. Mod. Lang. J. **79**, 90–99 (1995a)
14. MacIntyre, P.D., Gardner, R.C.: The subtle effects of language anxiety on cognitive processing in the second language [J]. Lang. Learn. **44**, 283–305 (1994)
15. Xian, Z., Guanyin, Z.: Analyzing and validating the construct of foreign language listening anxiety scale [J]. Modern foreign. Language. **34**(02), 162–170+219 (2011)
16. Shi, Y., Weiwei, F.: A study on college students' listening anxiety under the condition of compound dictation of CET-4 [J]. J. PLA Univ. Foreign Lang. **36**(06), 55–60+125–126 (2013)

Chapter 2
Research on the Reform of the Teaching Mode of "the Integration of Industry, Competition and Education" in Application-Oriented Universities in China: Take the Tourism Management Major of Shanghai S University as an Example

Jing Yin and Yuan Zhang

Abstract This study constructs a teaching evaluation system of "the integration of industry, competition and education" in application-oriented universities. Under the guidance of this system, taking the Tourism Management Major of Shanghai S University as an example, the paper explores the teaching reform mode of "three aspects and six transformations". The practice proves that this teaching mode plays a better role in promoting the cultivation of application-oriented tourism talents.

2.1 Introduction

At present, China's higher education has stepped into the popularization stage, and most of the undergraduate universities, with distinct application-oriented characteristics, are oriented to the local economy and society [1]. In 2022, the Ministry of Education of the People's Republic of China announced 147 double first-class universities (world-class universities and disciplines). By June 2023, there are a total of 1275 undergraduate universities in China, indicating that application-oriented undergraduate universities account for a large proportion. At the same time, with the transformation and adjustment of China's economic structure and the acceleration of industrial upgrading, the relationship between talent supply and the demand for economic and industrial development has undergone profound changes, and the country's demand for applied talents is becoming increasingly intense [2].

J. Yin (✉) · Y. Zhang
Department of Tourism Management, Shanghai Sanda University, Shanghai, China
e-mail: jying@sandau.edu.cn

S.-P. Tseng et al. (eds.), *Advances in Intelligent Information Hiding and Multimedia Signal Processing, Volume 2*, Smart Innovation, Systems and Technologies 416, https://doi.org/10.1007/978-981-97-8760-9_2

The integration of industry and education is an important direction for the high-quality development of China's higher education, especially in application-oriented universities, and has become an important policy of China's higher education in the new century.

The teaching mode of "the integration of industry, competition and education" refers to the industry demand as the guidance, teaching as the main body, and discipline competition as the link to promote the deeper and wider integration of industry and education, combine the discipline competition content with the professional ability, and promote the cultivation of high-tech applied talents.

2.2 Related Researches on the "the Integration of Industry, Competition and Education"

Taking "application-oriented undergraduate" and "industry, competition and education" as keywords searched on China CNKI, there are only 3 documents, which is relatively few. With the expanding search scope "industry, competition and education" as the keywords searched on China CNKI, a total of 85 documents exist. Among them, the earliest 2 documents were published in 2016, 16 from 2017 to 2019, 19 in 2020, 22 in 2021 and 26 in 2022. It can be seen that the subject of "the integration of industry, competition and education" has gradually attracted the researchers' attention in the past three years.

Through sorting out the documents, it is found that the research theme of the integration of industry, competition and education focuses on theoretical research, talent training mode, professional construction, course construction and other aspects. First, the theoretical research on the integration of "industry, competition and education" mainly includes sorting out the social background and development orientation formed by the university skills competition, explaining the current situation of the integration of industry and education in China's university skills competition, and discussing the value, function and significance of this model. Second, under the background of the integration of "industry, competition and education", some scholars have carried out studies about the talent training mode and put forward some innovative modes. For example, Yin Zhen [3] pointed out the multi-dimensional integration of "education, science, industry, innovation competition", and Zhu Jun [4] proposed the "new business" talent training mode integrating "industry, education, research, innovation, service and competition" based on the integration of industry and education ecosystem. Third, on the professional construction of the integration of "industry, competition and education", scholars mainly concentrate on resources integration, training base construction, faculty group construction, innovation and entrepreneurship, etc. [5–7]. Fourth, the research on the curriculum construction of the integration of "industry, competition and education" is career-oriented, mainly based on the needs of enterprises and industries and combined with the actual industry and discipline competition, as well as the teaching problems existing in the curriculum, and puts forward the reform idea of "industry, competition and education" integration [8, 9].

In summary, when it comes to the number of the documents, researches on the integration of industry, competition and education are few, but it has shown an upward trend in the past three years, and such topic has gradually become a research hotspot. Among them, there are relatively few studies on the teaching reform of tourism management majors in application-oriented undergraduate universities from the perspective of the integration of industry, competition and education. In terms of the contents, the research on the integration of industry, competition and education mainly focuses on theoretical research, talent training mode, professional construction, curriculum construction and other aspects. The research on the integrated teaching mode of industry, competition and education still needs to be deepened.

2.3 Construction of the Teaching Evaluation System of "the Integration of Industry, Competition and Education"

By studying the documents of the evaluation index system of teaching reform, combining with the interview of professional teachers, students and industrial tutors, the overall framework of the teaching evaluation system of "the integration of industry, competition and education" in application-oriented universities is preliminarily constructed, as shown in Table 2.1.

2.4 Empirical Study on "the Integration of Industry and Education" in Tourism Management Major of Shanghai S University

2.4.1 Basic Information of Tourism Management Major of Shanghai S University

Shanghai S University, founded in 1992, is a full-time general undergraduate university approved by the Ministry of Education. It is the first private undergraduate university in Shanghai. The university is positioned to build a multi-disciplinary, international and high-level university of applied technology, cultivating "reliable and practical" applied talents who can go farther.

The Tourism Management major of S University was established in 1992, and in 2020, it was approved as "Shanghai First-class Undergraduate Major Construction Site". Based on the local situations, the major adheres to the talent training mode of combining industry with education, uphold the new liberal arts construction concept of "combining expertise and integrating knowledge and action", and

Table 2.1 Teaching evaluation system of "the integration of industry, competition and education" in application-oriented universities

Primary indicators	Secondary indicators	Tertiary indicators
Career field	Jobs	Job cognition, job task, professional ability
	Job vocational certificate	"1 + X" certificate and related certificate passing rate
Competition field	Competition organization	Competition group, competition project, competition form
	Competition results	Number of awards, award level
	Teachers and students' sense of fulfillment	Students' personal growth and teachers' growth
Curriculum field	Teaching content	Curriculum group construction, teaching objectives, teaching tasks
	Teaching method	Teaching attitude, teaching guidance, assessment methods
	Teaching team	Industry mentor participation, double professional teachers
Integration degree	Production and class integration	The integration of post and class, class and certificate
	Class competition integration industry and competition integration	The connection between the competition group and the course group and the content of the course competition
		The combination of post and match ratio and match certificate connection

cultivate application-oriented innovative tourism talents to meet the new technology situation and the needs of the integrated development of culture and tourism. Since 2020, the Tourism Management major of Shanghai S University has successively organized students to participate in discipline competitions related to the first-class undergraduate construction project of Shanghai and the industry-university cooperation and collaborative education project of the Ministry of Education.

2.4.2 Teaching Reform Plan and Implementation Method of "the Integration of Industry, Competition and Education"

In recent years, the Tourism Management major of Shanghai S University is based on the application-oriented universities' teaching evaluation system of "the integration of industry, competition and education", and takes the discipline competition as the link, promotes the deep cross-integration of industry, competition and education, and constructs the teaching reform mode of "three aspects and six transformations" of collaborative education of industry, competition and education.

"Three aspects" refers to the three aspects of industry, competition and education, namely occupation (job cognition–vocational certificate–vocational ability) education, competition (competition group–competition project–competition ability) education, curriculum (curriculum group–curriculum content–knowledge and skills) education.

"Six transformation" refers to the implementation methods and means of cross-integration and transformation among the vocational education, competition education and curriculum education, namely, teaching, competition, project, curriculum, certificate and team.

2.4.2.1 Teaching: Based on the OBE Concept, the Job Requirements Are Transformed into Teaching Contents

On the basis of the OBE concept, visit the school-enterprise cooperative units to conduct research, understand the needs of employers, and investigate the job setting, job tasks, ability needs and career development path of related enterprises. At the same time, the industry experts are introduced to set up a professional construction committee composed of backbone teachers in the school and experts from outside the school, to jointly revise the training objectives of tourism management talents in our school. Thus, it determines the vocational ability goals that students should have, transform the vocational ability needs into teaching goals, and better clarify the theoretical and practical knowledge that students should be equipped with.

2.4.2.2 Competition: According to the Talent Training Objectives, Construct the Selection, Analysis and Group of the Competition

First, select the suitable competition. By analyzing the social or industrial recognition of the competition organizers, the influence and versatility of the competition, and comparing the training specifications and ability requirements in the professional talent training program, we, guided by the professional requirements, finally selected 6 competitions that are suitable for the students' future jobs. Second, analyze the selected 6 competitions one by one, carefully interpret the content of each competition, and clarify the tasks of the competition work and the demanding ability to participate in the competition. Third, according to the characteristics and requirements of the selected competition analysis, form an advanced ranking, and construct the discipline competition group of the tourism management major in our school.

2.4.2.3 Project: Build a Chain Progressive Competition Course Docking Course Group, and Connect the Course Module with Subject Competition Items

According to the structure system of the subject competition group, the corresponding 6 courses are selected to construct the competition course group. The course content of each stage is cross-integrated with the competition, and the progressive relationship with the course and competition is a chain, so as to ensure the orderly docking and comprehensive and deep integration at the competition and the curriculum level.

2.4.2.4 Curriculum: Reconstruct the Docking Course Content System and Innovate Teaching Methods

First, reconstruct the docking course content. Decompose the subject competition tasks of this course, and embed the needs of enterprises, the core work content of the job, and the evaluation points of each link of the discipline competition into the teaching content of the docking course. Second, innovate the hybrid teaching methods. Introduce project-based teaching according to the competition requirements, and carry out the mixed learning mode of in-class guidance and extracurricular competition practice. Third, establish a capability-oriented evaluation mechanism.

Change the traditional learning evaluation method, combine the needs of industrial development, subject competition projects, and establish an ability-oriented, process examination-oriented learning evaluation mechanism.

2.4.2.5 Certificate: Promote the Work of X Certificate and Mass Entrepreneurship and Innovation via Meeting the Needs of the Industry

Some discipline competitions of tourism management major are led by well-known enterprises in the industry, and the deepening assessment of professional skills is integrated in the process of the competition. Through promoting the industry competition, strengthen the students' intuitive and profound understanding of the industry vocational skills, and provide practical experience for the students to obtain the examination of the relevant industry vocational skills certificates, so as to promote the examination of vocational skills certificates. At the same time, we will further cultivate and incubate high-value entrepreneurship projects to further promote the innovation and entrepreneurship of students majoring in Tourism Management.

2.4.2.6 Team: Strengthen the Win-Win Mechanism and Build Up an Integrated Faculty Group

Introduce industry experts, strengthen the construction of "double-qualified" faculty group, and build up an integrated teaching team integrating interdisciplinary faculty group, cross-post enterprise mentor team and cross-professional competition mentor team. Encourage the teaching team to deeply carry out the integration of industry, competition and education, conduct practical exchange activities such as competition, teaching and research, improve the overall professional ability of the teaching team, transform the teaching results, and promote industrial development.

2.4.3 *Application Effectiveness*

At present, the teaching reform, based on "the integration of industry, competition and education", has been implemented consecutively for three years. In August 2023, a survey of the application effectiveness of the integration of industry, competition and education for tourism management students of 2019, 2020 and 2021 was carried out. A total of 160 questionnaires were collected, among which 158 points were valid questionnaires, with an effective rate of 98.75% (Table 2.2).

2.4.3.1 Students' Awareness of the Professional Field Has Improved and the Number of Professional Certificates Has Increased

Students' satisfaction score for the index of job analysis is 4.3, and their satisfaction score for the achievement of career certificate is 4.51. Compared with the teaching before the teaching reform, it can be found that students have a clearer cognition of the position and work tasks, a clearer understanding of their professional ability in the post, and a clearer learning goal. At the same time, the integration of curriculum certificates effectively improves the acquisition rate of students' vocational qualification certificates. During 2021–2022, Tourism Management students have obtained 201 industry certificates, with 1.7 certificates per capita.

2.4.3.2 Students Are Satisfied with the Competition Organization, which Have Achieved Certain Competition Results, and Thus Enhanced their Sense of Fulfillment

Students' satisfaction with the organization of the competition is 4.38 points. The students think that the competition group is reasonable and the selected competition items have high gold content, but the organization form of the competition can be further optimized. Students' satisfaction with the competition results is 4.59 points, and the number of students winning the competition has been greatly improved.

Table 2.2 Survey on the application effectiveness of industry, competition and education integration

Primary indicators	Secondary indicators	Tertiary indicators	5	4	3	2	1	Average score
		Job cognition	64	70	22	2	0	4.24
	Jobs	Vocational task	72	74	10	2	0	4.37
		Vocational ability	68	72	16	2	0	4.30
Career field			Subtotal					4.30
	Job vocational	Certificate acquisition	88	62	8	0	0	4.51
	Certificate							
		Total						4.35
	Competition organization	Competition group	86	60	10	2	0	4.46
		Competition project	96	50	10	2	0	4.52
		Competition form	60	66	30	0	2	4.15
		Subtotal						4.38
Competition field	Competition results	Number of awards, award level	98	58	0	2	0	4.59
	Students' sense of fulfillment	Students' sense of fulfillment	88	66	2	2	0	4.52
		Total						4.45
		Curriculum group construction	82	68	6	2	0	4.46
	Teaching content	Teaching objectives	84	66	6	2	0	4.47
		Teaching tasks	90	60	6	2	0	4.51
			Subtotal					4.48
		Teaching attitude	90	62	4	2	0	4.52
	Teaching method	Teaching guidance	80	64	12	2	0	4.41
		Assessment methods	80	68	8	2	0	4.43
			Subtotal					4.45
Curriculum field		Theoretical knowledge	74	70	12	2	0	4.37
		Job skills	78	70	8	2	0	4.42
	Teaching effect	Comprehensive quality of students	76	72	8	2	0	4.41
			Subtotal					4.40
		Industry Mentor participation	88	62	6	2	0	4.49
	Teaching team	Double professional teachers	76	72	8	2	0	4.41
			Subtotal					4.45
		Total						4.44
		The integration of post and class	70	76	10	0	2	4.34
	Production and class integration	The integration of class and certificate	72	72	12	0	2	4.34
			Subtotal					4.34
		The connection between the competition group and the course group	82	68	8	0	0	4.47

Table 2.2 (Continued)

Integration degree	Class competition integration	The cohesive content of the course and competition	88	68	2	0	0	4.54
			Subtotal					4.51
	Industry and competition integration	The combination of post and match ratio	60	88	6	2	2	4.28
		Match certificate connection	68	74	14	0	2	4.30
			Subtotal					4.29
		Total						4.38

During 2020–2022, students majoring in tourism management have won 48 awards in competitions at or above the municipal level, with a total of 142 students. The proportion of students winning the awards has increased from 3% to 29%. Students' satisfaction with the gain during the competition was 4.52 points, and the students generally said that they had improved their ability during the competition. At the same time, in the discipline competition, the teacher team members have also won the title of "Excellent Instructor" for many times, and the recognized teachers' enthusiasm for teaching is high.

2.4.3.3 Students Are Satisfied with the Teaching Organization after the Reform, and Their Comprehensive Quality Is Improved

The comprehensive rating of the course teaching level is 4.44 points, and the overall satisfaction with the course setting, course content, course teaching method, course teaching effect, and the configuration of the teacher team is high. Through the reform course teaching, the comprehensive quality of students has been effectively improved, students' ability to analyze and solve problems has been cultivated, and students' expression ability and teamwork consciousness have been enhanced. After entering the professional position, students can quickly transform the theoretical knowledge into the post practice needs, successfully complete the role change from student to employee, won the unanimous praise of employers.

2.4.3.4 Students Show the Highest Satisfaction to the Integration of Competition and Education, But the Integration of Industry and Competition Needs to Be further Strengthened

In terms of the cross-integration of the three levels of industry, competition and education, students show a high satisfaction to the integration of competition and education, reaching 4.51 points. It is generally believed that the course and the competition content are closely connected and have a high degree of integration.

The evaluation of the integration of industry and class is 4.34 points. The students think that the connection between post and class is scientific and reasonable, and the integration of class certificates is strong. The satisfaction score of industry-competition integration is 4.29 points. Students believe that the proportion of post-competition integration and the way of competition and certificate connection need to be further optimized.

2.5 Conclusion

In this study, through studying relevant literature and combining with interviews with experts, the teaching evaluation system of "the integration of industry, competition and education" in application-oriented universities is constructed. Under the guidance of this system, taking the tourism management major of Shanghai S University as an example, the teaching reform mode of "three aspects and six transformations" is industry demand oriented, teaching as the main body, and discipline competition as the link. The practice has proved that this teaching mode plays a better role in promoting the cultivation of the application-oriented tourism talents. Through the above path exploration, we hope we can provide some references for the construction of tourism management major in application-oriented undergraduate universities.

Fund Support The key teaching reform project of Shanghai Shanda University, Research on the Teaching Mode Reform of "the Integration of Industry, Competition and Education" of Tourism Management Major in Application-oriented Undergraduate Universities (No.: A020203.23.001.02).

References

1. Fang Jinghua, F.: Study on the Value of the Integration of Industry and Education to the Training of Application-Oriented Undergraduate Talents and its Realization Path. Guangdong Polytechnic Normal University (2022)
2. Wang Liya, F.: Multiple measures to promote the innovation of talent training mechanism for application-oriented undergraduate universities. Shandong Education. **17**, 32–33 (2021)
3. Yin Zhen, F., Cao Ziyang, S., Bo Jie, T.: Research and practice of multi-dimensional integration of "teaching, science, industry and innovation competition" for mechanical composite talents training model. Theoretical Res. Pract. Innov. Entrepreneurship. **4**(16), 124–126 (2021)
4. Zhu Jun, F.: Research on the integrated talent training mode of "industry, learning, research, innovation, service and competition" based on the integration of industry and education ecosystem. Time-Honored Brand Marketing. **23**, 175–178 (2022)
5. Liu Jianqian, F., Yu Ziying, S., Lu Yanqin, T.: Teaching dilemma and countermeasures of curriculum group from the perspective of the integration of industry, competition and education——a case study of fashion design major in Jiaxing University. J. Jiaxing Univ. **35**(01), 139–144 (2023)

6. Zhou Li, F.: Research on curriculum teaching reform based on OBE education concept—a case study of tourism planning and development curriculum teaching reform. J. Xinjiang Inst. Educ. **35**(02), 10–14 (2019)
7. Chu Tiantian, F., Chen Rui, S.: Research on the promotion mechanism of "industry, teaching and research competition" to the training mode of applied talents in marketing majors. Sci. Heilongjiang. **8**(13), 80–81 (2017)
8. Mao kehong, F.: Exploration and practice of big data professional talent training mode based on the "integration of industry, education and competition". Comput. Knowl. Technol. **17**(11), 115–117 (2021)
9. Zhong Huizhen, F.: An analysis on the path of combining competition and teaching in private universities under the background of new liberal arts – A case study of logistics management major in S universities in Jiangsu province. Reform Opening Up. **(24)**, 48–53 (2021). https://doi.org/10.16653/j.cnki.32-1034/f.2021.024.007

Chapter 3
Design and Implementation of Outdoor Navigation Robot System Based on Google Map API and Improved YOLOv7-tiny

Chien-I Chang, Din Yuen Chan, An-Chao Tsai, Jhing-Fa Wang, and Jie-Wen Du

Abstract To suppress the world-wise impact on the manpower of a sudden epidemic pandemic, the demands of inexpensive autonomous robots for outdoor deliveries of food, medicines, and pollution will have fast growth. Traditional autonomous robots rely on time-consuming offline SLAM mapping, constraining their adaptability and applicability in large-range outdoor environments. Moreover, varying fields require repetitive map construction. Outdoor settings introduce mobile obstacles, further reducing SLAM mapping navigation system's practicality and stability. To address these critical problems, this paper presents an innovative non-SLAM mapping outdoor navigation technology, enhancing outdoor navigation system stability while significantly reducing costs. This cost-effective navigation system holds promise for outdoor robots, providing both affordability and deployment flexibility. The next stage for developing our robot is attentive to more substantial benefits on societal safety and healthcare.

3.1 Introduction

Recent robotics advancements, especially in autonomous mobile robots (AMRs), have spurred interest in outdoor navigation. AMRs find application in logistics,

C.-I. Chang (✉) · D. Y. Chan
National Chiayi University, Chiayi, Taiwan
e-mail: dychan@mail.ncyu.edu.tw

A.-C. Tsai
National PingTung University, PingTung, Taiwan
e-mail: actsai@mail.nptu.edu.tw

J.-F. Wang · J.-W. Du
National Cheng-Kung University, Tainan, Taiwan
e-mail: wangjf@csie.ncku.edu.tw

S.-P. Tseng et al. (eds.), *Advances in Intelligent Information Hiding and Multimedia Signal Processing, Volume 2*, Smart Innovation, Systems and Technologies 416, https://doi.org/10.1007/978-981-97-8760-9_3

manufacturing, healthcare, and services. Traditional SLAM mapping using radar scanning faces limitations like inaccurate obstacle detection, especially glass, and time-consuming mapping. Also, dynamic obstacles hinder SLAM-based outdoor navigation.

This study tackles these issues by focusing on Non-SLAM mapping. It merges computer vision and GPS to improve outdoor navigation stability. Leveraging Google Maps API and upgraded YOLOv7-tiny algorithm, real-time data aids route planning and object recognition. This research aims to create an efficient outdoor navigation robot adept at dynamic environments. Subsequent chapters detail related work, implementation, results, and conclusions, revealing the outdoor navigation robot's design with Google Maps API and improved YOLOv7-tiny.

3.2 Proposed System

3.2.1 System Overview

Figure 3.1 shows the entire hardware system architecture.

Figure 3.2 shows the architecture of the entire navigation system.

3.2.2 Hardware

The hardware equipment is the outdoor robot CMICSD (Cognitive Multimedia Integrated Circuit System Design Lab) completed in this paper by [1]. The CMICSD tracked vehicle is shown in Fig. 3.3a, b shows the overall hardware circuit diagram.

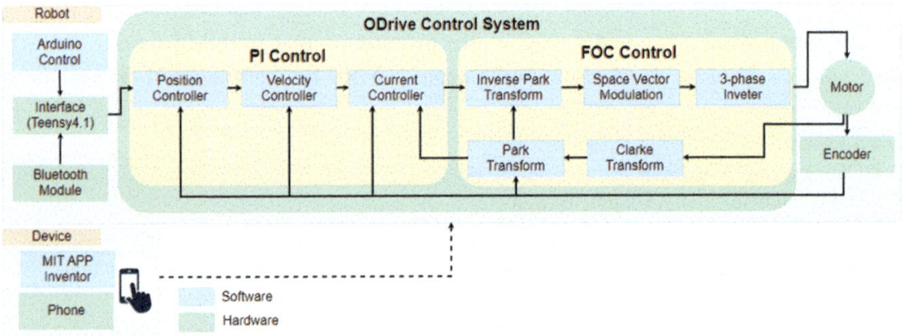

Fig. 3.1 The hardware system overview

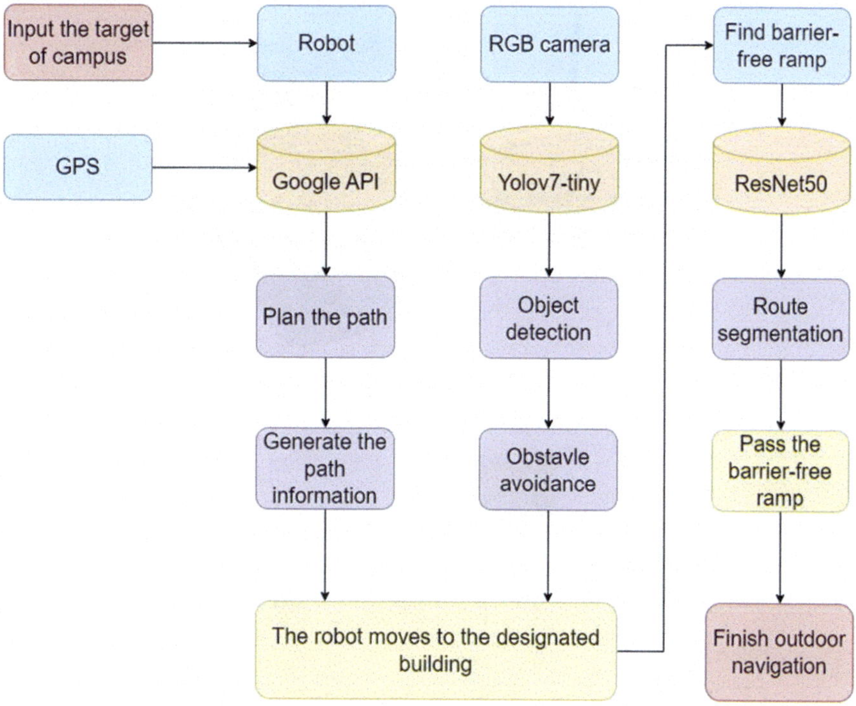

Fig. 3.2 The hardware system overview

3.2.3 *Control System*

The architecture of the Odrive control system is shown in Fig. 3.4, and the more detailed architecture diagram is shown in Fig. 3.5, where the structure of this control system is a FOC combined with PI control.

3.2.4 *Navigation User Interface*

The front-end system plays a critical role in providing a user-friendly interface for controlling and monitoring the robot's navigation. The architecture is shown in Fig. 3.6.

(a)

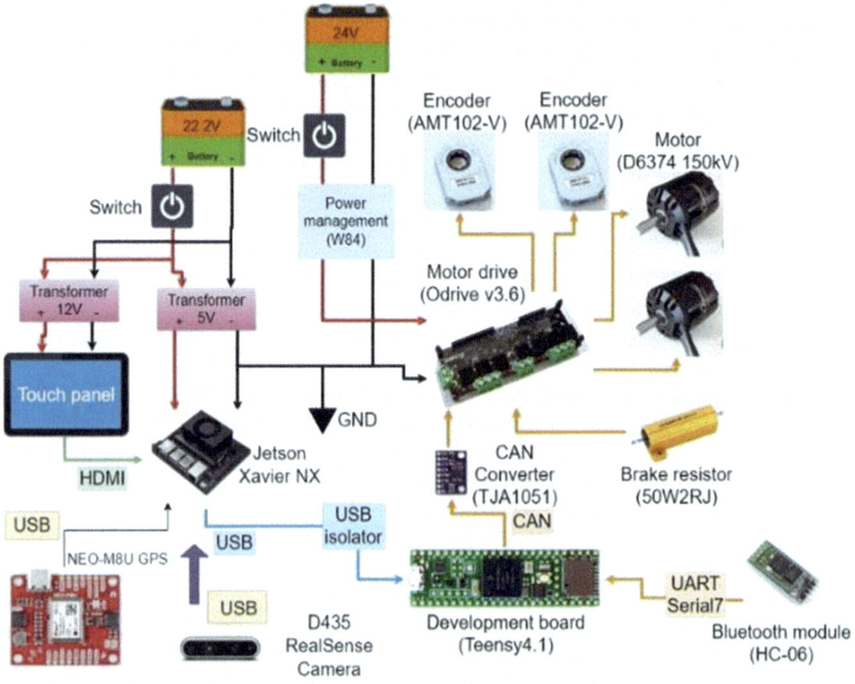

(b)

Fig. 3.3 (**a**) The CMICSD tracked vehicle. (**b**) The hardware circuit diagram

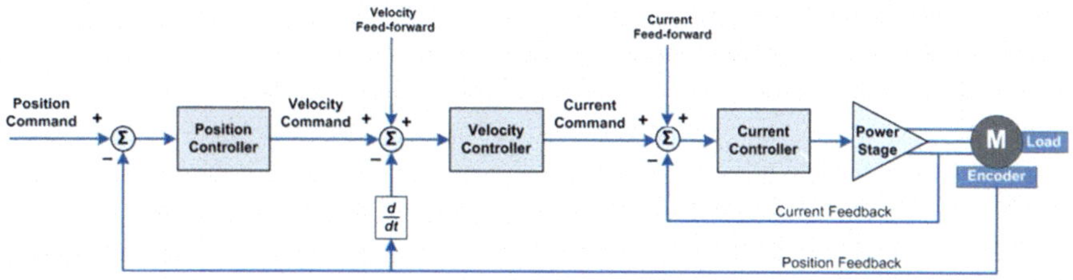

Fig. 3.4 Odrive system architecture

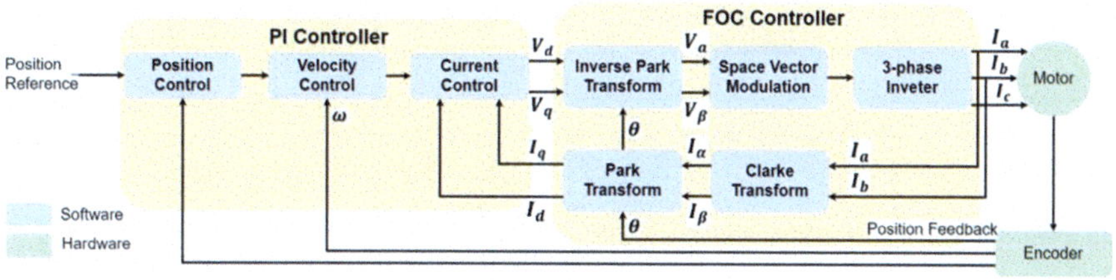

Fig. 3.5 FOC combined with PI control system architecture

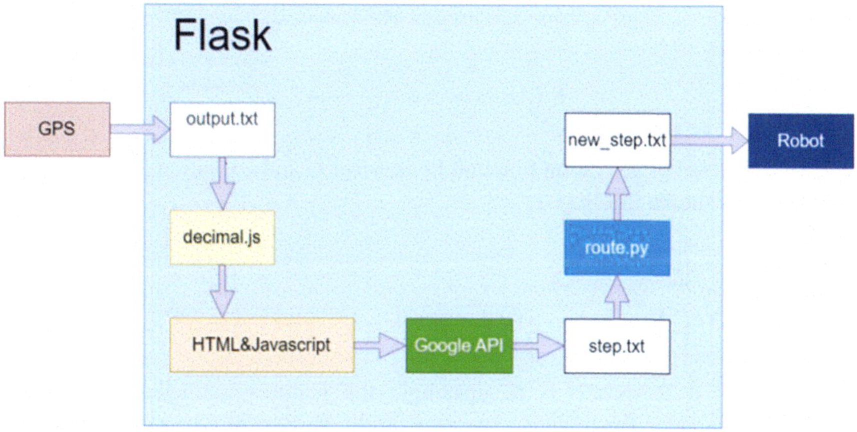

Fig. 3.6 Front-end system architecture diagram

3.2.5 *Improved Yolov7-tiny*

In the outdoor navigation system, YOLOv7-tiny can be enhanced and customized to improve its performance. The activation function LeakeyReLU is replaced by the activation function SiLU. Activation function LeakeyReLU and the formulas of the activation function SiLU are as follows:

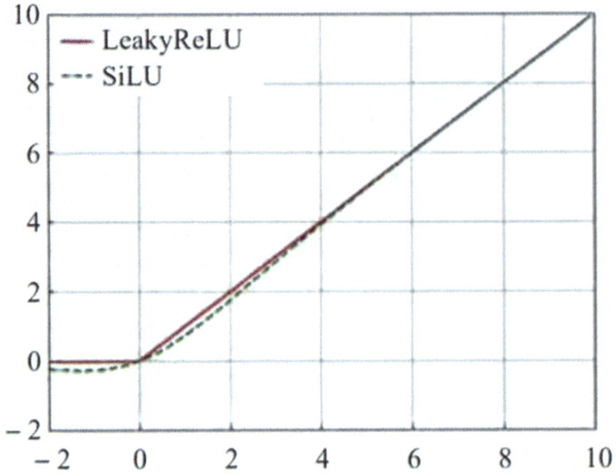

Fig. 3.7 Comparison of curve smoothness between LeakyReLU and SiLU

$$\text{LeakeyReLU}(x) = \begin{cases} x, x > 0 \\ ax, x \leq 0 \end{cases}$$

$$\text{SiLU}(x) = x \times \frac{1}{1 + e^{-x}} \tag{3.1}$$

The curve level of activation function LeakeyReLU and activation function SiLU comparison is shown in Fig. 3.7.

3.2.6 *ResNet50*

The Decoder of ResNet50 is to upsample the features extracted and restore image size. Encoder for feature extraction is the ResNet50 Decoder that adopts PPM + Deepsup (Pyramid Pooling Module + Deep supervision trick). The process is shown in the Decoder deep supervision network architecture as shown in Fig. 3.8. The database is taken from MIT ADE20K [2]. It can be used for scene perception, segmentation, analysis, multi-object recognition and semantic understanding.

3.2.7 *Outdoor Navigation Algorithm*

The outdoor navigation algorithm is a crucial component of the outdoor navigation robot system. It is responsible for processing sensor data, incorporating localization information, and generating the navigation commands necessary for the robot

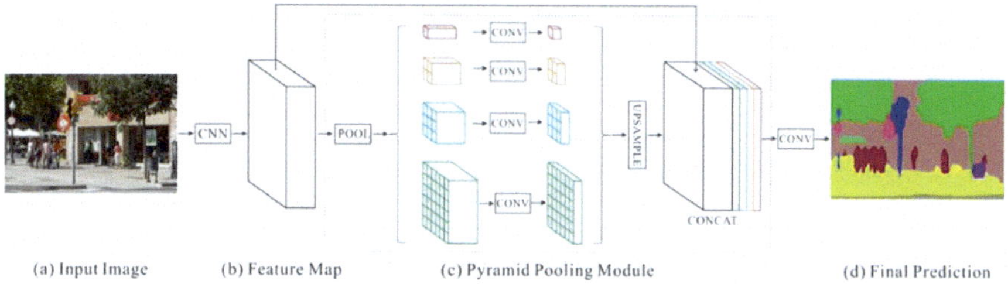

Fig. 3.8 PPM+ deep supervision network architecture diagram

to navigate autonomously in outdoor environments. The algorithm combines the Google Maps API and the improved YOLOv7-tiny object detection algorithm to achieve accurate and efficient navigation. The outdoor navigation algorithm consists of Sensor Data Processing, Path Planning, Object Detection and Recognition, Navigation Command Generation, Feedback and Adaptation. System not only needs to use a GPS module and RealSense Camera hardware equipment but also needs to use a lot of software to complete together. Fig. 3.9a shows the flow chart of this navigation system, and Fig. 3.9b shows the avoidance algorithm Obstacle Avoidance Flowchart for safely navigating outdoors to avoid obstacles. Because the GPS module will be affected by factors such as satellites, weather, clouds, etc., resulting in inaccurate and unstable positioning, this outdoor navigation system will store the route map when it knows the current location and destination and can be used in When the GPS cannot receive satellite signals, walk according to the route map stored at the beginning to realize more humanized navigation.

3.3 Experimental Results

3.3.1 Experiment of Navigation UI Interface

Figure 3.10 shows the overall navigation user interface, including a search box, a field for displaying the latitude and longitude of the current location, a field for manually entering the current latitude and longitude, a field for the latitude and longitude of the destination, and a visual map display.

3.3.2 Experiment of Yolov7-tiny and Avoidance

By fusing the results of Yolov7 object recognition and the depth map of the depth camera, the depth information, recognition results, and categories are displayed at the same time, so that the robot can know the obstacle category and depth, and

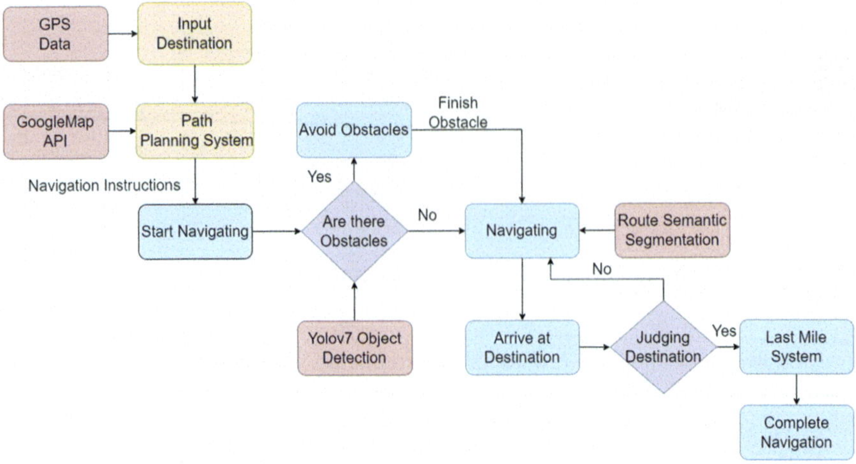

(a)

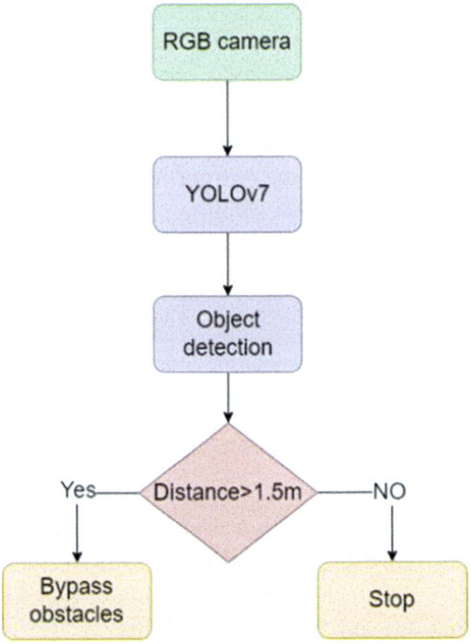

(b)

Fig. 3.9 (**a**) Flow chart of this navigation system. (**b**) Obstacle avoidance flowchart

realize the stability and accuracy of obstacle avoidance. We further experimented with the impact of incorporating the ResNet50 model into an outdoor navigation system. The function of route semantic segmentation enables the robot to better understand the surrounding environment and convert the image into a binary image

Fig. 3.10 Navigation UI

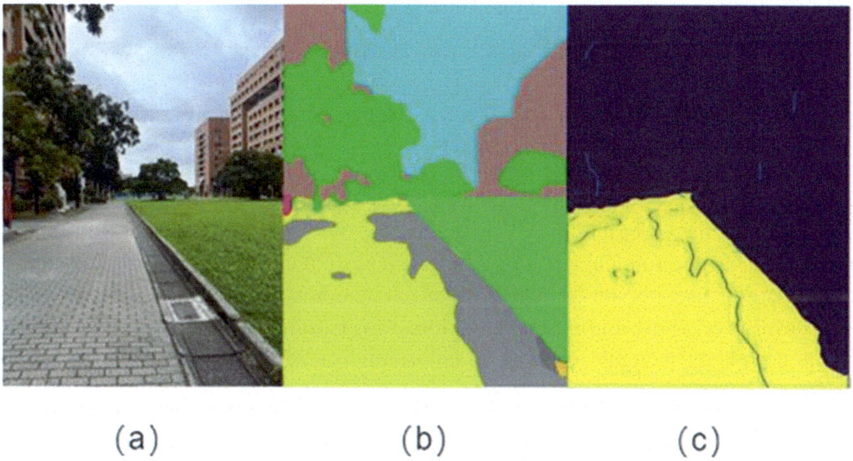

Fig. 3.11 (**a**) Actual image, (**b**) Image after segmentation, (**c**) Post-processed image

through post-processing so that the robot can easily judge the walkable and non-walkable areas and make wise navigational decision-making, Fig. 3.11 shows the actual image, image after semantic segmentation and post-processed image.

3.3.3 Experiment of Mobile and Outdoor Navigation

We evaluated the motion performance and navigation accuracy of the robot through multiple adjustments of parameters and outdoor field experiments. We evaluated the outdoor navigation system, and the robot can locate its location through GPS. Through the self-written user interface and GoogleMap API, users can input the destination they want to go to, and then output the route information to the robot so that the robot can move according to the route information, while moving, use Yolov7-tiny to carry out depth information with the depth camera for object detection and obstacle avoidance, and through the route semantic segmentation module, try to avoid places where the robot cannot walk. When arriving at the destination, it will confirm whether the destination given by the GoogleMap API is the same as the current location. If it is consistent, it will navigate the barrier-free ramp. With the route semantic segmentation module, it will pass the barrier-free ramp smoothly, and then the instructions of the robot's current navigation will be recorded to make the error of the next navigation smaller. After continuous error correction and field testing, the results show that the robot can smoothly walk between the various buildings in the Ziqiang Campus.

3.4 Conclusion

In this study, we propose the design and application of an outdoor navigation robot that combines the GoogleMap API with the improved YOLOv7-tiny algorithm and implements an economical outdoor navigation system created by [1] tracked robot, implementing a set of Non-SLAM outdoor navigation system. The goal of this work is to develop a robust and efficient outdoor navigation system that enables robots to autonomously navigate in outdoor environments while detecting and avoiding obstacles.

References

1. Research design and application of self-developed outdoor mobile robot with FOC combined with PI control. Available: https://hdl.handle.net/11296/7f7gb7
2. ADE20K. Available: https://groups.csail.mit.edu/vision/datasets/ADE20K/

Chapter 4
Integrating DeepLabCut and ChatGPT for Dolphins Skeleton Detection and Behavior Analysis

Shao-En Hsu and Jhing-Fa Wang

Abstract This chapter aims to streamline the analysis of long-duration dolphin videos while reducing the workload of dolphin caretakers. To address this challenge, we utilized the capabilities of DeepLabCut, an open-source toolbox designed for pose estimation, and extended its capabilities through advanced animal assembly and tracking methods, which are particularly crucial for scenarios involving multiple dolphins. It is worth noting that there is limited research focusing on behavioral analysis of dolphins at present. Dolphin behavior holds significant ecological and academic implications; however, traditional observation techniques are time-consuming and often imprecise. To tackle this issue, we proposed an innovative approach that combines DeepLabCut with ChatGPT for rapid analysis. We annotated key skeletal points on dolphins and subsequently trained a model using a pretrained ResNet neural network. Postprocessing the model's output for body parts allows us to swiftly obtain data on dolphin behaviors. Finally, this behavior data serves as ChatGPT's database, enabling long-term analysis of dolphin behavior patterns. Our method not only ensures precise marine behavior analysis but also contributes substantially to the conservation and research of dolphins. Experimental results demonstrate that this system achieves an impressive accuracy rate of 92.5% and can effectively detect both the skeleton and behavior of dolphins.

4.1 Introduction

Dolphins, as highly intelligent marine creatures, have always been a focal point in ecology, ethology, and conservation studies [1]. Their behavioral patterns and interactions not only shed light on the health status of marine ecosystems but also offer deep insights into social dynamics and learning abilities. However, prolonged video observations, often coupled with manual annotations, demand

S.-E. Hsu (✉) · J.-F. Wang
Department of Electrical Engineering, National Cheng-Kung University, Tainan, Taiwan

S.-P. Tseng et al. (eds.), *Advances in Intelligent Information Hiding and Multimedia Signal Processing, Volume 2*, Smart Innovation, Systems and Technologies 416, https://doi.org/10.1007/978-981-97-8760-9_4

substantial time, resources, and are inevitably susceptible to human biases, possibly resulting in imprecise or inconsistent interpretations. In the context of modern technological advancements, automated image analyses and machine learning offer a fresh approach to these concerns.

This research endeavors to address these challenges by leveraging the synergy of deep learning and natural language processing (NLP) techniques. Specifically, our goal is to develop a streamlined and effective methodology for the analysis of long-duration dolphin videos, ultimately reducing the workload of dedicated dolphin caretakers and contributing to a more comprehensive understanding of dolphin behavior. To achieve this, we turn to DeepLabCut [2–4], an open-source toolbox renowned for its prowess in pose estimation. Building upon its existing capabilities, we enhance it with advanced animal assembly and tracking techniques, essential for scenarios involving multiple dolphins interacting in their natural or captive environments.

In addition to simplifying the analysis process, we increase the accuracy and objectivity of behavioral observations. Traditional methods often rely on manual annotation and interpretation, thereby introducing potential biases and inconsistencies. We use a pretrained ResNet neural network to automatically identify key skeletal points of dolphins. Then, the postprocessing method is used to analyze the key skeleton points predicted by the model, so that we can analyze the behavior of dolphins. Finally, the analyzed results are sent to ChatGPT as a database, so that caretakers can inquire about the status of dolphins through the ChatGPT UI web developed in this study. The main objectives of this research can be summarized as follows: (1) Automatic tracking of dolphin skeletons using DeepLabCut. (2) Use postprocessing to conduct behavioral analysis and classification on the generated body coordinates. (3) Use the detected behavior as ChatGPT's database to give it the ability to answer questions about dolphin conditions. (4) Automate behavioral analysis to shorten long videos and reduce workload for caregivers.

4.2 Related Work

4.2.1 Dolphin Behavior and Research

Deep learning methodologies have transformed the landscape of animal pose estimation in recent years. A significant stride in this domain has been the development of DeepLabCut [2–4]. Mathis et al. [2] pioneered this framework, showing its applicability in various settings and advocating its accuracy in recognizing distinct body parts without traditional markers. The investigation delved into the feature detection architecture of DeeperCut, a noteworthy pose estimation algorithm. It highlighted that even with a relatively small number of training images (approximately 200), human-level labeling accuracy can be achieved. This achievement can be primarily attributed to the practice of transfer learning, which is based on deep neural networks pretrained on ImageNet [5], a vast object recognition dataset. Furthermore, they highlighted that end-to-end training can amplify the performance of the network. Using this method, by labeling only a few hundred frames, they were able to train specialized feature detectors that could accurately identify body parts.

Building on this groundwork, Nath et al. [3] expanded the horizons of DeepLabCut by illustrating its competency in 3D markerless pose estimation [3]. Their work emphasized the tool's versatility, showcasing its effectiveness across different species and a spectrum of behaviors. Such 3D capabilities are pivotal in understanding intricate dynamics, especially in environments with depth variations, an aspect crucial for aquatic animals like dolphins.

However, studies involving multiple animals present a unique set of challenges, particularly in distinguishing and tracking multiple subjects concurrently. Lauer et al. [4] tackled this challenge, enhancing DeepLabCut's capabilities to accurately estimate poses, identify, and track multiple animals simultaneously [4]. They adapted ImageNet-pretrained ResNets, EfficientNets, and DLCRNet-ms5. They trained the network to predict output from input images using annotated key points for target score maps, location refinement maps, and PAFs.

Given these advances, current research aims to use the advantages of DeepLabCut to conduct behavioral analysis of animals, such as dogs [6] and monkeys [7]. However, there is currently no literature analyzing dolphins, so this study wants to start with dolphins.

4.2.2 ChatGPT

With the evolution of Natural Language Processing (NLP) technologies [8], chatbots have transitioned from basic query-response systems to highly intelligent assistants capable of profound interactions and comprehension. OpenAI's GPT model series has demonstrated outstanding performance in natural language generation, especially when it comes to parsing and generating contextually relevant text. Given the complexity of dolphin behaviors, employing GPT models to analyze and generate coherent linguistic descriptions can offer deeper insights into their actions. Based on the description in Leo Wang's article [9], utilizing GPT models combined with Google's TTS technology can create a multilingual chatbot. Applying this technology to dolphin behavior logs enables caretakers to understand the conditions of dolphins in real time, rather than traditional manual review and analysis of logs. This not only offers caretakers an intuitive and efficient interactive platform but also provides a rich data resource for future behavioral studies.

4.3 Proposed System

4.3.1 System Overview

This thesis presents the system architecture in its entirety, depicted in Fig. 4.1. Three web cameras are installed at approximately 5 meters above a water pool to capture daytime footage of dolphins. This setup is aimed at enhancing the training dataset. Firstly, the dolphin images are extracted at a rate of one frame every 5 seconds. Label the points of interest on extracted frames, including rostrum, melon, blowhole, dorsal fin, pectoral fins, bodypart1-3, and tail fin. Train the DeepLabCut system with annotated images. The backbone of the neural network we choose is ResNet-50 [5]. Compared with other neural networks, ResNet50 has a faster analysis speed. This process generates xy-coordinate files (CSV files) for each individual at various time points. Using these coordinate files, behavioral analysis is performed, resulting in textual descriptions of dolphins' behaviors at different time intervals. Ultimately, this data is utilized as a database for ChatGPT. As a result, dolphin caretakers can use ChatGPT to acquire information about the dolphins' conditions and behaviors.

4.3.2 Dataset Description

In partnership with Farglory Oceanpark, we obtained a video dataset spanning 6 hours and meticulously annotated it. This dataset comprises a total of 4,320 RGB

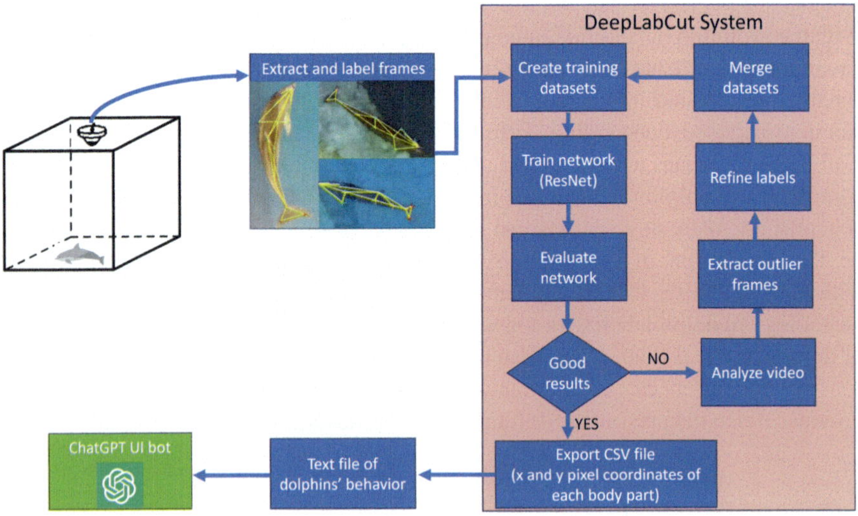

Fig. 4.1 Overall block diagram

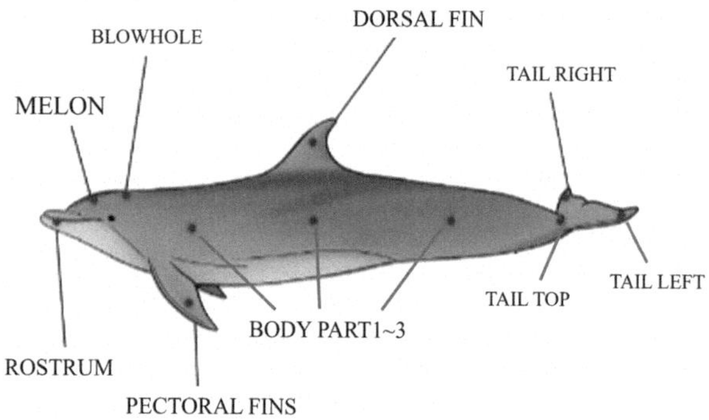

Fig. 4.2 Dolphin's features

images capturing multiple dolphins in various poses. We divided this dataset into three distinct subsets: 80% designated for training, 3% for validation, and 17% for testing, ensuring robust model evaluation. Each image has 11 features with manual annotations, which will be taught to the model. Dolphin's features are shown in Fig. 4.2.

4.3.3 Architecture of Network

The DeepLabCut utilizes a 50-layer ResNet architecture, with training weights based on ResNet as shown in Fig. 4.3. The architecture and parameter settings are followed in our implementation.

To evaluate the model's performance, we used root-mean-squared error (RMSE) as the evaluation metric. RMSE measures labeling accuracy in pixel units, and its formula is shown in Eq. (4.6).

$$RMSE = \sqrt{\frac{1}{n}\sum_{i=1}^{n}(y_p - y_l)^2} \tag{4.1}$$

Here, "yp" represents the predicted value, and "yl" represents the labeled value, providing a quantitative measure of labeling accuracy.

We inputted 1280 × 720 RGB images into the system and encoded the positional details of 11 characteristics for each image. The output of the system consists of a 1280 × 720 × 11 score map, which displays activation levels for each pixel in the image that corresponds to each body part. The processing pipeline initiates with the input being directed to a convolutional layer (Conv1) equipped with 64 7×7 filters, batch normalization, and a subsequent 3×3 max-pooling operation. Subsequently, the input image passes through several layers of convolution blocks and identity blocks, as shown in Fig. 4.3. Following an extensive training regimen comprising

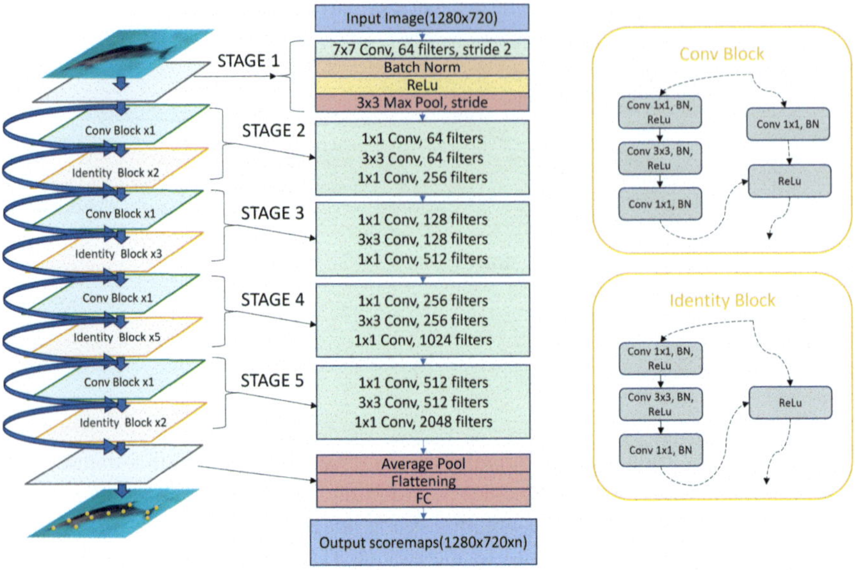

Fig. 4.3 Network architecture

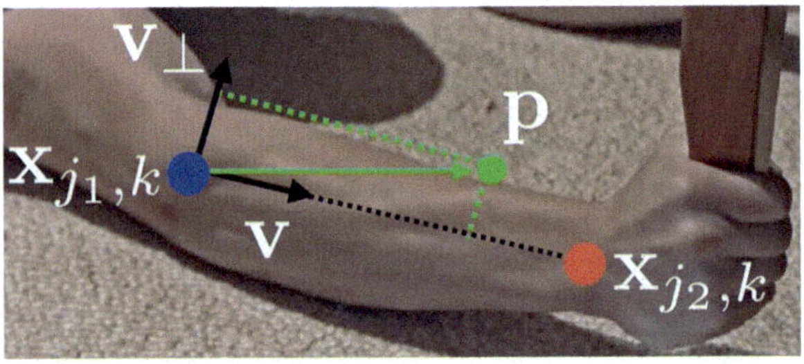

Fig. 4.4 Definition of PAF [10]

32,000 iterations, the network yields a comprehensive set of outcomes, including Part Affinity Fields (PAF), score maps, and the precise coordinates of each feature's points.

PAF stands for Part Affinity Fields and was initially introduced by Openpose [10] as a technology for human posture estimation. PAF is a vector field that operates in two dimensions and characterizes the correlation between various parts of the image, as illustrated in Fig. 4.4.

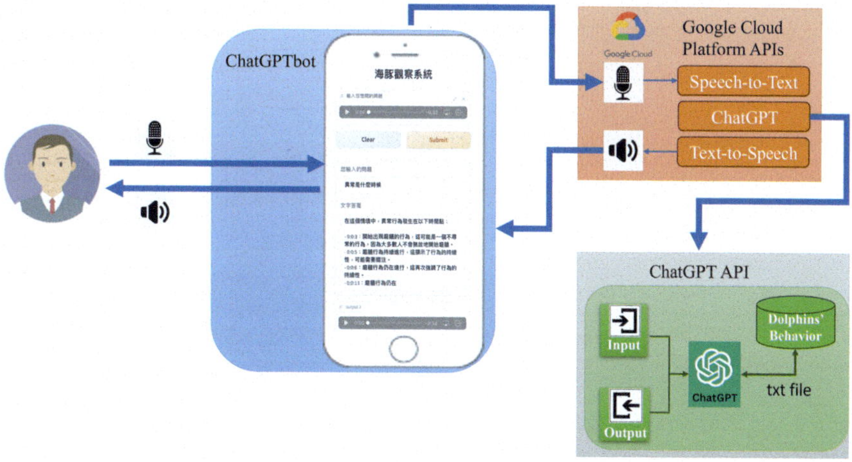

Fig. 4.5 System overview of ChatGPT UI Web

4.3.4 ChatGPT UI Web

The system overview of ChatGPT UI Web is shown in Fig. 4.5. After analyzing the coordinates points obtained from the network, we can extract insights about the dolphins' behaviors at different time points. This data is then utilized as a database for ChatGPT. The system is integrated with Gradio, allowing seamless connection with Google Cloud and ChatGPT. This integration enables the system's accessibility across various devices, enabling users to input their queries and receive answers effortlessly. The database encompasses information about instances when the dolphins engage in behaviors such as wall rubbing and fish spitting, along with corresponding timestamps. This compilation proves advantageous for dolphin care personnel, providing them with valuable insights for observation and monitoring purposes.

4.4 Experimental Results

In the dolphin behavior analysis experiments, we subjected a video of 5 hours and 19 minutes to analysis, using it as the test dataset. The resolution of the video was 1280 × 720 and it had a frame rate of 15 fps. The analysis process took a total of 4 hours and 32 minutes to finish. Furthermore, we used another video of 1 hour and 43 minutes as the training dataset. The configured parameters are as follows: the batch size is 8. For the first 7,500 iterations, the learning rate remained at 0.0001; for iterations 7,500 to 12,000, the learning rate was adjusted to 5e-05, and subsequently, for the remaining iterations, it was fine-tuned to 1e-05. We then compared the RMSE

Fig. 4.6 RMSE values and confidence scores for each body part

RMSE Values and Confidence Scores Per Feature for 32k Iteration		
Dolphin Feature	RMSE	confidence score
Rostrum	11.21	91.8%
Melon	6.72	93.3%
Blowhole	8.61	92.5%
Dorsal Fin	12.63	89.4%
Pectoral Fin	14.26	83.1%
bodypart1	9.78	87.3%
bodypart2	9.24	91.7%
bodypart3	8.69	95.3%
Tail top	7.00	98.3%
Tail left	9.26	94.5%
Tail right	7.87	96.2%

Experimental Environment		
Hardware	Operating System	Windows
	CPU	Intel i7-12700K
	GPU	Nvidia RTX3060
	RAM	32GB
Software	Python3.8+TensorFlow2.10.0+Deeplabcut2.2.2	

Fig. 4.7 Experimental environment

values and confidence scores between various body parts. The summarized results are presented in Fig. 4.6.

Figure 4.7 outlines the hardware and software environment of the system. We employed TensorFlow 2.10.0 to ensure compatibility with DeepLabCut2.2.2. To expedite the training process, we used a Nvidia RTX 3060 graphics card, which significantly enhanced the training speed.

In Fig. 4.8, the feature prediction and location points of the dolphin are displayed. The manually labeled points are represented by "X" and the model-predicted points are represented by ".". Additionally, we analyzed the PAF maps and score maps output by the final layer, as shown in Fig. 4.9. It is evident from the PAF map and score map that the network accurately predicts dolphin skeletons.

We also plotted a 2D behavioral trajectory plot for the dolphins, shown in Fig. 4.10. We analyzed the coordinates file of the final output parts. If the dolphin's body stays at the wall edge for more than 3 seconds, it is classified as wall rubbing. If the tail fin is vertically above the head for more than 5 seconds, it is classified as fish spitting. We extracted the timestamps of these behaviors and exported them to text files. Then, we inputted text files into our ChatGPT UI web system. This allows us to access the timestamps of abnormal behaviors at any time, as shown in Fig. 4.11.

During the final testing phase, our system's ability to detect dolphin behavior was found to be highly accurate. We analyzed 31 minutes (1860 seconds) of video footage and obtained the following results: (1) The system accurately identified

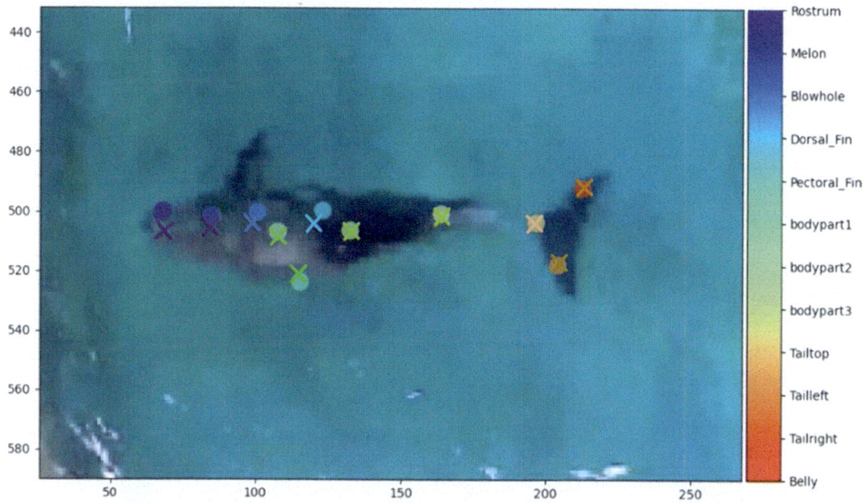

Fig. 4.8 The feature prediction and location points of the dolphin

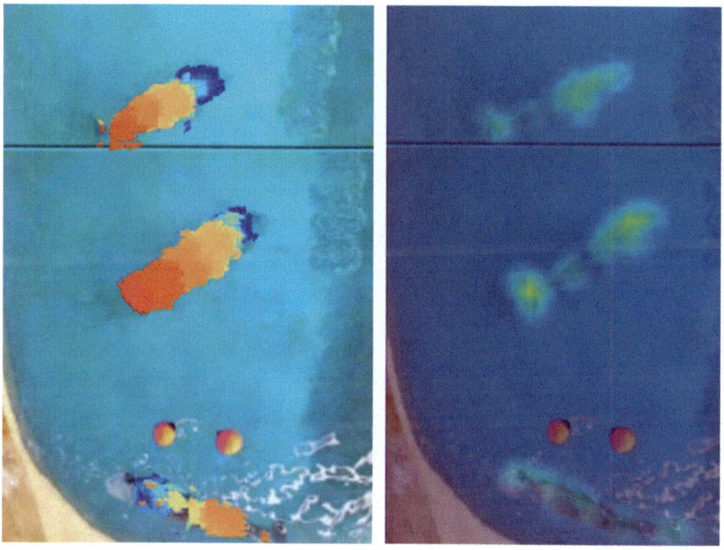

Fig. 4.9 The score map (right) and PAF map (left) of the dolphins

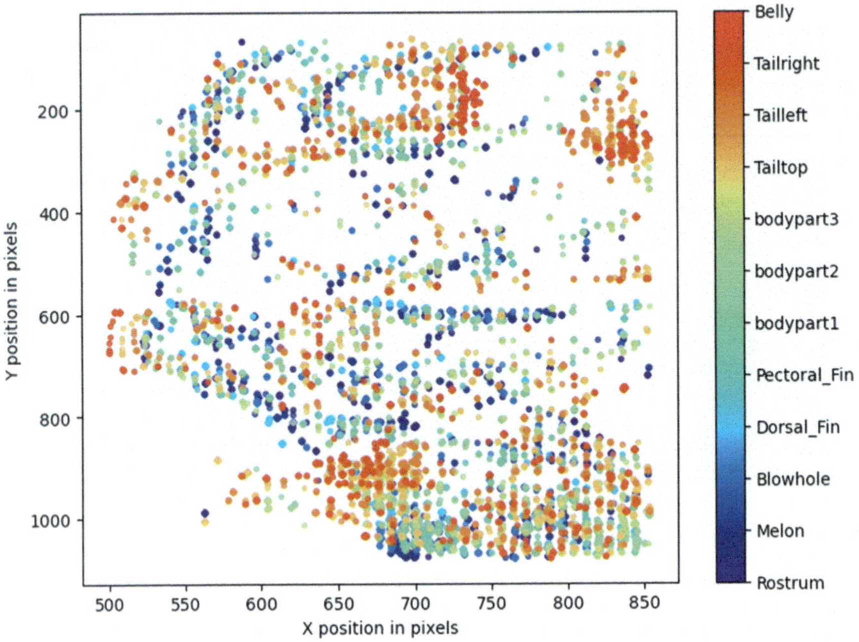

Fig. 4.10 The behavioral trajectory plot of the dolphin

Fig. 4.11 Confusion matrix of our result

	"Actual" Positive	"Actual" Negative
"Predicted" Positive	True Positive(912)	False Positive (32)
"Predicted" Negative	False Negative(106)	True Negative(810)

instances of dolphin skeletons and their behavior for 912 seconds. (2) There were 32 seconds where the system incorrectly marked nonexistent dolphin skeletons or behavior. (3) The system failed to identify dolphins in 106 seconds where they were present. (4) For 810 seconds, there were no dolphins or detection of their skeletons or behavior (Fig. 4.12).

$$\text{Accuracy} = \frac{(TN + TP)}{(TN + FN + FP + TP)} = 92.5\% \tag{4.2}$$

$$\text{Precision} = \frac{TruePositive}{(TruePositive + FalsePositive)} = 96.6\% \tag{4.3}$$

$$\text{Recall} = \frac{TruePositive}{(TruePositive + FalseNegative)} = 89.5\% \tag{4.4}$$

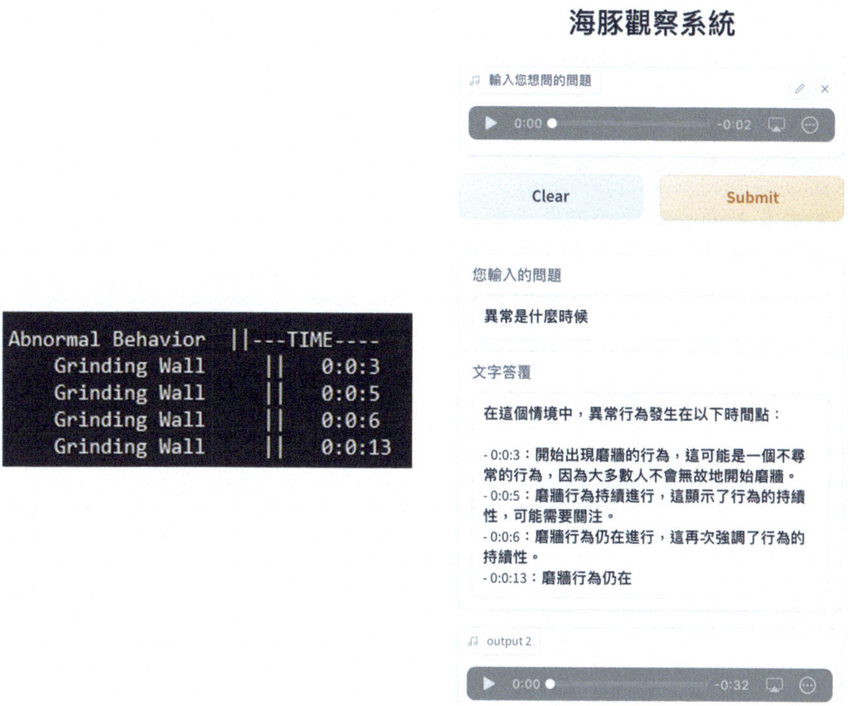

Fig. 4.12 The behavior result and ChatGPT result

4.5 Conclusion

In this chapter, we utilized DeepLabCut to annotate thousands of images for training purposes. Through training, the model was enabled to accurately identify and learn the features of dolphins. Subsequently, we employed postprocessing techniques to analyze the model's predictions regarding the location of these features, enabling us to determine the behaviors of the dolphins. Finally, the obtained results were input into ChatGPT's database. Caregivers can access this information by using the ChatGPT UI web interface to inquire about the current status of the dolphins.

We developed a new system that combines DeepLabCut and ChatGPT technologies to identify dolphins' skeletons and behaviors. The system employs ChatGPT to analyze the collected data. With a success rate of 92.5%, the system has proven to be highly accurate in detecting dolphins' skeletons and behaviors. Additionally, it has helped in reducing the length of several-hour videos by half, which will significantly ease the burden on dolphin caretakers who previously had to conduct prolonged observations.

References

1. Shane, S.H., Wells, R.S., Würsig, B.: Ecology, behavior and social organization of the bottlenose dolphin: a review. Mar. Mamm. Sci. **2**(1), 34–63 (1986)
2. Mathis, A., Mamidanna, P., Cury, K.M., Abe, T., Murthy, V.N., Mathis, M.W., Bethge, M.: Deeplabcut: markerless pose estimation of user-defined body parts with deep learning. Nat. Neurosci. **21**(9), 1281–1289 (2018)
3. Nath, T., Mathis, A., Chen, A.C., Patel, A., Bethge, M., Mathis, M.W.: Using deeplabcut for 3d markerless pose estimation across species and behaviors. Nat. Protoc. **14**(7), 2152–2176 (2019)
4. Lauer, J., Zhou, M., Ye, S., Menegas, W., Schneider, S., Nath, T., Rahman, M.M., Di Santo, V., Soberanes, D., Feng, G., et al.: Multi-animal pose estimation, identification and tracking with deeplabcut. Nat. Methods **19**(4), 496–504 (2022)
5. He, K., Zhang, X., Ren, S., Sun, J.: Deep residual learning for image recognition. In: Proceedings of the IEEE Conference on Computer Vision and Pattern Recognition (CVPR) (2016)
6. Ferres, K., Schloesser, T., Gloor, P.A.: Predicting dog emotions based on posture analysis using deeplabcut. Future Internet **14**(4), 97 (2022)
7. Labuguen, R., Bardeloza, D.K., Negrete, S.B., Matsumoto, J., Inoue, K., Shibata, T.: Primate markerless pose estimation and movement analysis using deeplabcut. In: 2019 Joint 8th International Conference on Informatics, Electronics & Vision (ICIEV) and 2019 3rd International Conference on Imaging, Vision & Pattern Recognition (icIVPR), pp. 297–300. IEEE, New York (2019)
8. Chowdhary, K.R.: Natural Language Processing, pp. 603–649. Springer, New Delhi (2020). https://doi.org/10.1007/978-81-322-3972-7_19
9. Build a talking ChatGPT Bot in any language with Google Text-to-speech (TTS) and OpenAI. https://www.linkedin.com/pulse/build-talking-chatgpt-bot-any-language-google-tts-openai-leo-wang
10. Cao, Z., Simon, T., Wei, S.-E., Sheikh, Y.: Realtime multi-person 2d pose estimation using part affinity fields. In: Proceedings of the IEEE Conference on Computer Vision and Pattern Recognition, pp. 7291–7299 (2017)

Chapter 5
Framework for Virtual Try-on with Background Processing and Human Pose Estimation

Chih-Chiang Lin and Jhing-Fa Wang

Abstract Virtual try-on systems have gained immense popularity, offering convenience and personalized experiences for online shopping. However, many existing systems lack robust handling of diverse human poses and background settings. This study proposes a comprehensive Framework for Virtual Try-On with Background Processing and Human Pose Estimation, leveraging both pre-trained and custom-trained models. The framework commences by utilizing U2Net (Qin, Zhang, Huang, Dehghan, Zaiane, Jagersand, Pattern Recogn, 106:107404, 2020) for background segmentation, preserving the original backdrop for later use. For the human subject, the system employs pre-trained OpenPose and DensePose models for 2D and 3D pose estimation, respectively. Further semantic segmentation is accomplished using a custom-trained Mask2Former model, identifying distinct body and clothing areas. The clothing item also undergoes similar segmentation. All these processed inputs are then fed into a pre-trained GP-VTON model for the final virtual try-on. The synthesized output is seamlessly integrated with the original background, providing a realistic virtual try-on experience. Experimental results confirm the effectiveness of our framework in handling a wide range of poses and backgrounds.

5.1 Introduction

With the rise of online shopping and virtual worlds, accurately previewing how clothes look without trying them on has become a significant research challenge. Traditional fitting experiences often require consumers to physically visit stores, which is not only time-consuming but also impractical in certain situations like global pandemics. Therefore, virtual try-on technologies have emerged, allowing users to quickly and accurately preview clothes on online platforms. However, existing virtual try-on solutions mostly work only in specific backgrounds and

C.-C. Lin (✉) · J.-F. Wang
Department of Electrical Engineering, National Cheng-Kung University, Tainan, Taiwan

S.-P. Tseng et al. (eds.), *Advances in Intelligent Information Hiding and Multimedia Signal Processing, Volume 2*, Smart Innovation, Systems and Technologies 416, https://doi.org/10.1007/978-981-97-8760-9_5

for specific body poses, lacking the flexibility and wide applicability needed. Thus, this research aims to develop a highly flexible and customizable virtual try-on framework that integrates background processing and human pose estimation. Our system employs U2Net for segmenting the human figure and its background and uses OpenPose [2] and DensePose [3] to predict 2D and 3D human poses, respectively. Additionally, we have custom-trained a semantic segmentation model based on Mask2Former [4], designed specifically for the fine-grained segmentation of human bodies and clothing. Lastly, these pre-processed data are fed into the GP-VTON virtual try-on model [5] for the synthesis and rendering of clothes. The main contributions of this paper are as follows: (1) Introduction of a comprehensive and flexible virtual try-on framework that can handle a variety of backgrounds and human poses. (2) Custom training of an efficient semantic segmentation model specifically designed for the fine-grained segmentation of human bodies and clothes. (3) Effective acceleration of the pre-processing steps for virtual try-on using pre-trained models.

5.2 Related Work

The development of virtual try-on technologies involves various subfields, including image segmentation, human pose preprocessing, and the actual algorithms employed for virtual try-on. Each of these areas has seen notable contributions, creating a rich tapestry of techniques that have influenced our research. We present a comprehensive review of these fields to establish the foundation for our study, provide comparisons with existing methods, and offer insights into future research directions.

5.2.1 Image Segmentation

Our methodology makes extensive use of U2-Net for image segmentation tasks. This network architecture serves as the foundation for our more advanced applications, such as apparel and human body parsing. Known for its high efficiency and flexibility, U2-Net has shown great promise in handling a variety of image segmentation tasks.

5.2.2 Human Pose Preprocessing

Among the notable contributions in human pose preprocessing is the work of Cao et al., who introduced part affinity fields (PAF). This efficient technique for real-time multi-person 2D pose estimation has a significant impact on enhancing both the performance and the accuracy of virtual try-on systems. Another advancement is DensePose, a method focused on dense human pose estimation across various backgrounds. It employs a novel technique that maps 2D RGB images to 3D human surfaces, thereby achieving highly accurate pose estimations. Furthermore, Mask2Former is capable of handling multiple image segmentation tasks such as panoramic, instance, and semantic segmentation. This framework extracts local features by limiting cross-attention within the predicted mask regions, making it highly suitable for precise human pose preprocessing.

5.2.3 *Virtual Try-on Techniques*

In the realm of virtual try-on solutions, VITON stands out as an image-based network that synthesizes realistic images by parsing human body representations. It even maintains its own database for 2D human portraits and apparel [6]. Diverging from the conventional requirement of paired training images, TryOnGAN utilizes generative adversarial networks (GANs) to learn the relationship between different individuals' apparel and poses. This approach results in a more flexible and efficient virtual try-on system [7]. Ge et al. simplified the virtual try-on process by using knowledge distillation, which eliminates the need for human feature maps while maintaining the quality of the synthesized output [8]. Last but not least, GP-VTON employs Local Warping techniques along with human body part parsing to offer an accurate and realistic virtual try-on experience [5].

Collectively, these studies and technologies constitute the theoretical and practical backbone of our research, providing a multi-faceted viewpoint and toolkit for the future development and application of virtual try-on technologies.

5.3 Proposed System

5.3.1 *System Overview*

In order to offer a highly accurate and user-friendly virtual try-on experience, we introduce a comprehensive virtual try-on system. The system's detailed architecture, covering both the inference and training phases, is vividly captured in Figs. 5.1 and 5.2, respectively. The framework mainly comprises the following integral components:

Background and Human Segmentation Using Pre-trained U2-Net: This step primarily focuses on accurately segmenting the human subject from the background in the given images. Human Pose Estimation: Use of Pre-trained OpenPose for 2D Pose Estimation: This aims to capture the 2D spatial orientation of the human body. Use of Pre-trained DensePose for 3D Pose Estimation: This is designed

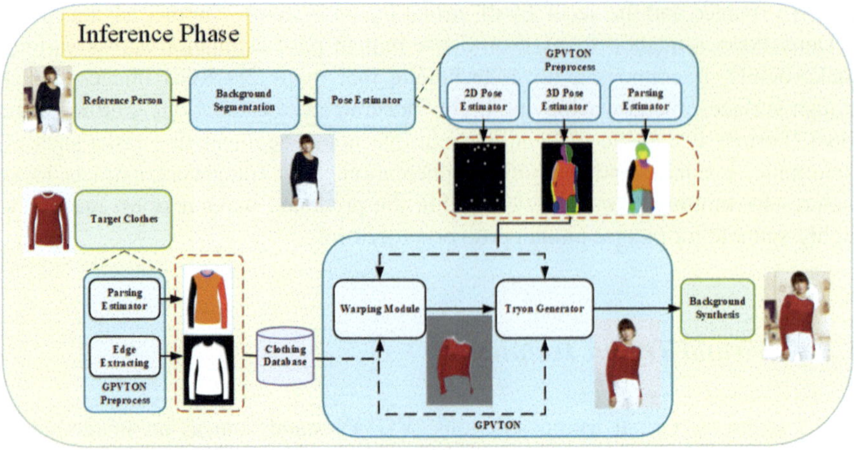

Fig. 5.1 System overview (inference phase)

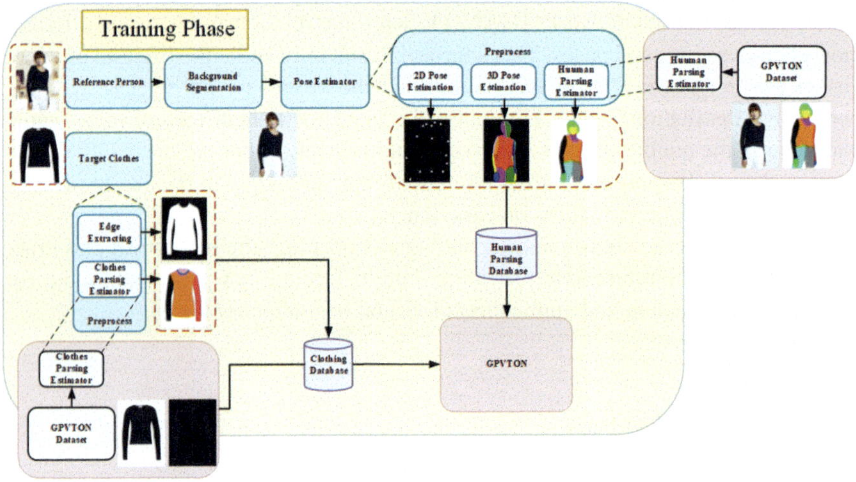

Fig. 5.2 System overview (training phase)

to acquire the 3D spatial orientation of the human body. Semantic Segmentation Using Mask2Former Architecture and Training: This step offers finer semantic segmentation of the human body to precisely capture its posture. Virtual Try-On Framework Using Pre-trained GP-VTON: This component takes the pre-segmented and pose-estimated human model and synthesizes it with clothing items meticulously. Background Reintegration Using Adaptive Masking and Inpainting: After the virtual try-on process, the segmented human model is reintegrated with the originally segmented background to produce a natural-looking virtual try-on output. Each of these components will be elaborated upon in the subsequent sections.

5.3.2 Background and Human Segmentation Using Pre-trained U2-Net

To achieve higher precision and efficiency in segmenting the background and human figures, we employ the pre-trained U2-Net model from the rembg library. This model excels at delivering high-accuracy foreground and background separation. Through this preliminary preprocessing step, we successfully isolate the human figure from the background, thereby providing a clean and accurate foundation for subsequent human pose estimation and virtual garment fitting procedures. The architecture of the model can be seen in Fig. 5.3.

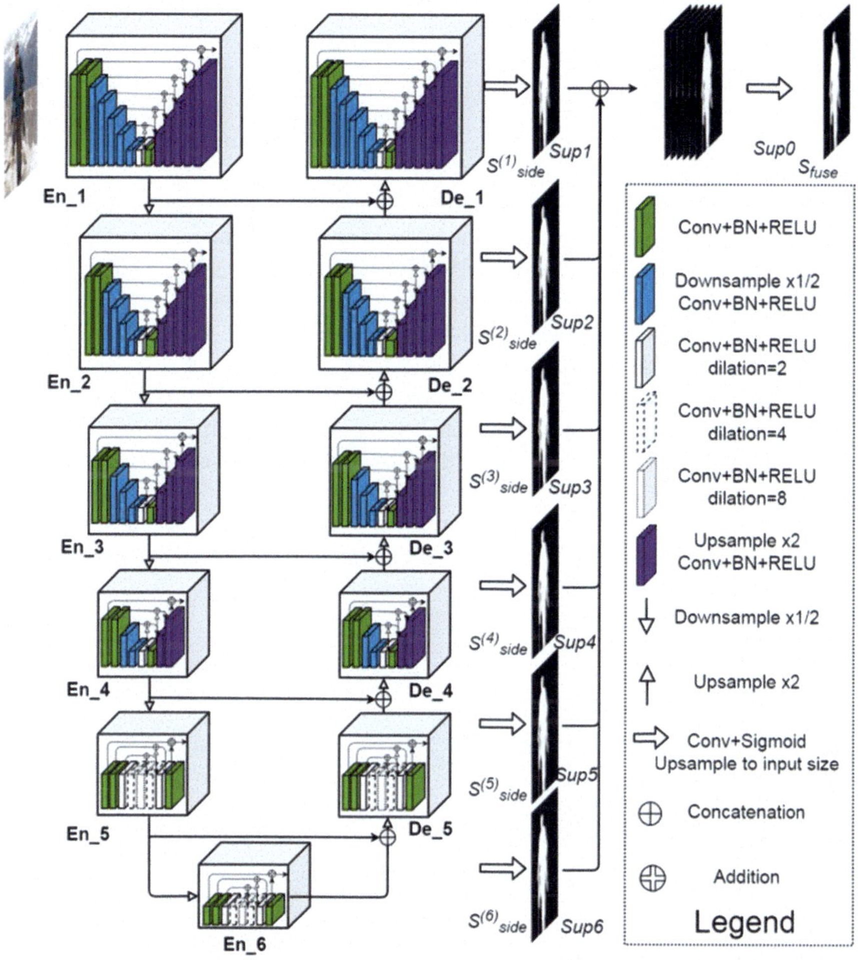

Fig. 5.3 The network architecture of u2net

5.3.3 Human Pose Estimation

Within the Human Pose Estimation section, we elaborate on three crucial elements integral to the performance and capability of our virtual try-on system. These components, although independent, collaboratively enhance the system's overall accuracy. Firstly, we make use of a pre-trained OpenPose model to estimate 2D poses. Secondly, a pre-trained DensePose model aids us in achieving 3D pose estimation. Lastly, unlike the aforementioned components that rely on pre-trained models, our system employs a semantic segmentation model that we have trained using Mask2Former architecture. These three elements are pivotal in capturing detailed human contours, understanding human postures in various dimensions, and precisely segmenting the human body into meaningful semantic regions. We will delve into the specifics, including their respective roles and contributions, in the sub-sections that follow.

5.3.3.1 Use of Pre-trained OpenPose for 2D Pose Estimation

In our virtual try-on system, the accuracy of human pose estimation is paramount. To achieve this, our system leverages the body 25 pre-trained model from OpenPose. This model is designed to identify 25 primary key points on the human body, covering essential regions like the head, upper limbs, and lower limbs. OpenPose operates on a multi-stage convolutional neural network (CNN). The first set of stages is focused on predicting part affinity fields (PAFs), which delineate the connectivity between various body parts. Subsequent stages generate confidence maps for each of these key points. These dual functionalities work in tandem to deliver accurate and actionable pose estimation. For a detailed look into the architecture, refer to Fig. 5.4.

The output key points follow the format [x, y, confidence], aligning seamlessly with the input requirements of our system's subsequent modules, ensuring smooth integration across the entire pipeline.

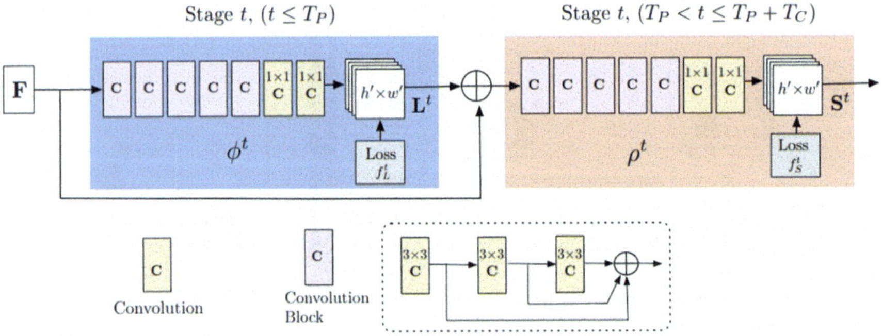

Fig. 5.4 The network architecture of multi-stage CNN

5.3.3.2 Use of Pre-trained DensePose for 3D Pose Estimation

3D pose estimation plays a crucial role in rendering realistic and accurate fittings. We employ DensePose-RCNN, developed by Facebook AI Research (FAIR), as our 3D Pose Estimator. This model is part of the larger DensePose toolkit and integrates region-based convolutional neural networks (R-CNN) with dense pose mapping capabilities.

The architecture of DensePose-RCNN can be categorized into three main components:

Object Detection: The model employs R-CNN to detect human figures in the given image. Region Extraction: Post-detection, it isolates regions corresponding to individual human figures. Dense Pose Estimation: Lastly, 3D pose estimation is applied to each extracted region to generate detailed mapping between 2D images and 3D body surfaces. For this application, we selected the DensePose-RCNN model based on ResNet-101 architecture due to its proven efficiency and high accuracy. This enables a more natural and believable try-on experience. A schematic of the DensePose-RCNN architecture can be seen in Fig. 5.5.

5.3.3.3 Semantic Segmentation Using Mask2Former

To achieve a more natural and authentic virtual try-on experience, we utilize the features of GP-VTON with its local-flow global-parsing (LFGP) module, which focuses on individual deformation of clothing across specific body parts such as the left arm, torso, and right arm. To precisely obtain these fine-grained part details, we employ the Mask2Former tool developed by Facebook AI Research (FAIR).

Mask2Former uses a cutting-edge transformer architecture and masked attention mechanism, specifically designed for high-accuracy semantic segmentation. In our application context, the model not only accurately segments the image into

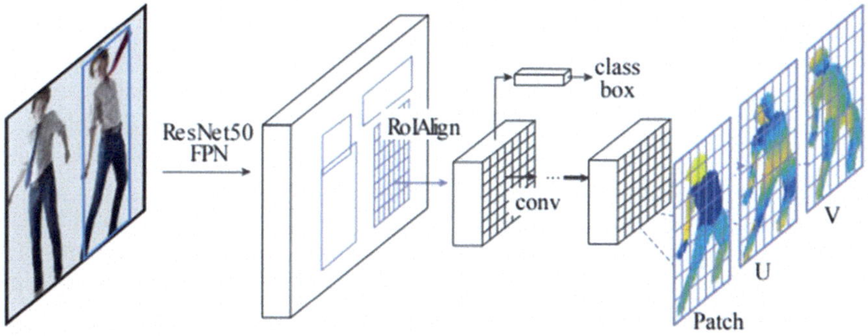

Fig. 5.5 The network architecture of DensePose-RCNN

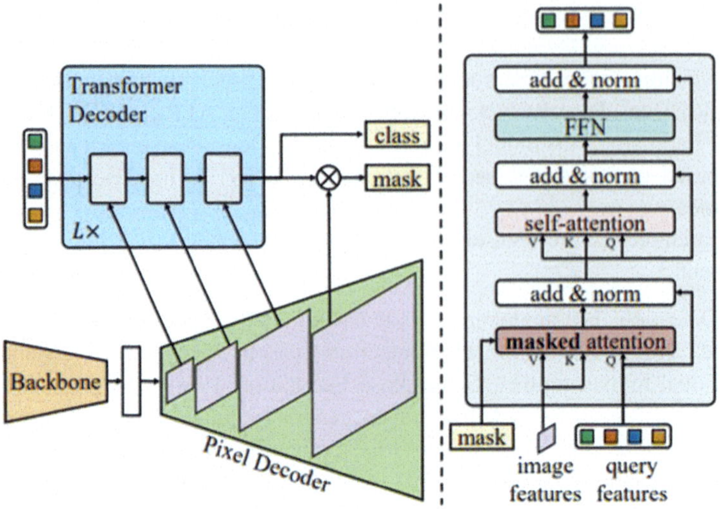

Fig. 5.6 Overview of Mask2former

background, hair, torso, etc., but also further subdivides it into cloth-covered and uncovered parts of each arm and leg, as well as shoes.

This level of detailed segmentation allows us greater flexibility and precision in subsequently deforming each portion of the clothing during the virtual try-on process. This results in a more natural and authentic try-on experience. For a detailed architecture and functionalities of the model, please refer to Fig. 5.6.

5.3.4 Virtual Try-on Framework Using Pre-trained GP-VTON

Traditional virtual try-on systems often rely on global warping modules. While generally effective, these modules can lead to synthesis failures when dealing with complex inputs such as diverse human poses or intricate clothing designs. This is because global warping struggles to preserve semantic information across different clothing regions, resulting in texture distortions. To address these challenges, we employ the GP-VTON model with its local-flow global-parsing (LFGP) deformation module. The LFGP approach utilizes localized flows specific to individual body parts for independent deformation of each clothing region. This enables the preservation of semantic information across all garment areas, even when dealing with complex inputs. The result is a highly accurate and semantically coherent virtual try-on experience, effectively mitigating common issues like texture distortion and semantic inconsistencies found in traditional methods. Figure 5.7 provides an architectural overview of the GP-VTON model.

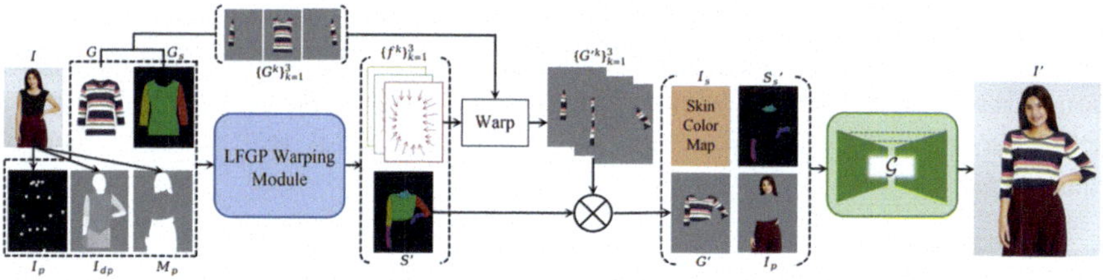

Fig. 5.7 Overview of GP-VTON

5.3.5 Background Reintegration Using Adaptive Masking and Inpainting

To enhance the realism and integration of the virtual try-on result with its original background, we've refined our approach for merging the segmented human figure after virtual garment fitting with the preserved background. Rather than a simple overlay, we harness an adaptive inpainting strategy to ensure seamless blending.

Initially, we perform background segmentation on the yet-to-be-virtually-dressed human figure. The segmented human figure and the separated background each possess transparency information. To identify the regions where inpainting is needed, we calculate a combined mask based on the transparency levels of both the human figure and the background.

Setting a threshold, e.g., 150, for the alpha channels:

$$\text{combinedalpha} = \text{logical and }(\text{bgalpha} < \text{threshold}, \text{human} < \text{threshold}) \tag{5.1}$$

This mask determines where the transparency of both the background and the human figure falls below the threshold. We then utilize this mask for the inpainting step.

Subsequently, we utilize the Telea inpainting algorithm based on the Fast Marching Method to process the background. This algorithm effectively blends the edges of the target area, providing a natural transition that eliminates any abrupt changes and enhances visual realism.

By leveraging this method, we ensure that the final virtual try-on image looks more cohesive and natural, avoiding the "cut-paste" appearance that can arise from simpler overlay methods.

5.4 Experimental Results

In this chapter, we will present the outcomes of each major module in the proposed system sequentially. These modules encompass background and human segmentation, 2D and 3D pose estimation, semantic segmentation using Mask2Former, virtual try-on synthesis based on pre-trained GP-VTON, and background reintegration. We will highlight how each module uniquely contributes to the overall virtual try-on process and how they collaborate to achieve a highly realistic and coherent final output.

5.4.1 Background and Human Segmentation

In this subsection, we focus on the role of background and human segmentation in our proposed system, which is achieved using the pre-trained U2-Net from the rembg library. This step is instrumental for two primary reasons: first, it provides a clean human figure required for the virtual try-on stage, and second, it separates the background for subsequent reintegration.

Figure 5.8 presents a comparison between the original input image and its segmented components: the human figure and the background. The accurate segmentation sets the stage for a successful virtual try-on process. After segmentation, the isolated human figure serves as the subject for virtual garment fitting. This ensures that the clothing items are correctly placed over the intended areas, avoiding any overlay on the background. Moreover, the segmented background is conserved for later use in the pipeline. During the background reintegration stage, this preserved background is used to seamlessly merge with the virtually dressed human figure, thus enhancing the visual realism of the final output.

5.4.2 2D Pose Estimation

In this subsection, we delve into the significance of 2D pose estimation in our virtual try-on system. Utilizing the pre-trained body 25 model from OpenPose, our system is capable of identifying 25 key points on the human figure.

Figure 5.9 showcases the result of this 2D pose estimation applied to our previously segmented human figure. As depicted, the 25 key points accurately capture the essential bodily structures, encompassing areas like the head, torso, arms, and legs. These points are essential to guide the correct positioning of virtual garments onto the human figure.

5.4.3 3D Pose Estimation

In this subsection, we focus on 3D pose estimation using DensePose-RCNN, which serves as an integral component in achieving realistic garment fitting in our system.

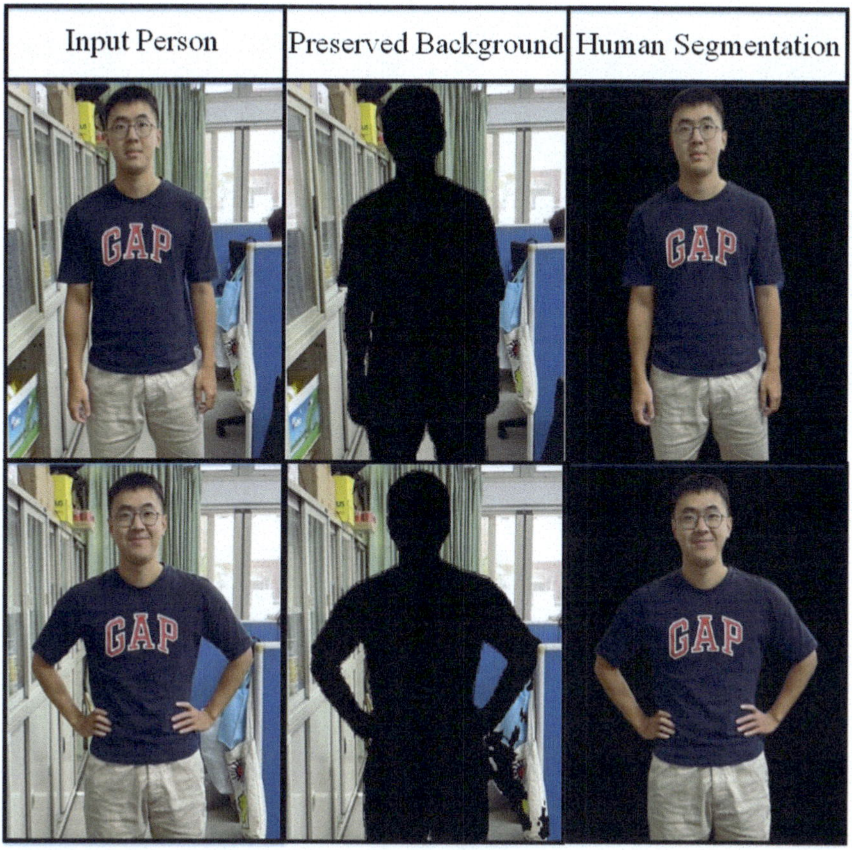

Fig. 5.8 Background and human segmentation

Figure 5.10 shows the original segmented human figure alongside its 3D pose estimation results. This estimation provides a detailed mapping between the 2D image and 3D body surfaces, allowing for a more accurate and realistic virtual try-on experience.

The DensePose-RCNN model successfully identifies and maps intricate body parts, contributing to the quality of the subsequent virtual try-on stages. This proves the efficacy of using DensePose-RCNN for 3D pose estimation in our system.

5.4.4 Mask2Former Segmentation: Body and Clothing Parts

The accuracy and granularity of semantic segmentation are key factors in achieving a natural and authentic garment fitting experience. Fig. 5.11 showcases the output of

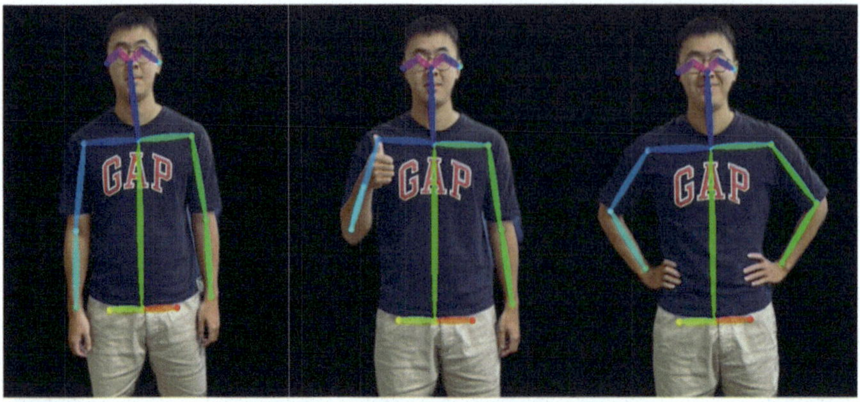

Fig. 5.9 Key point detection on segmented human subject

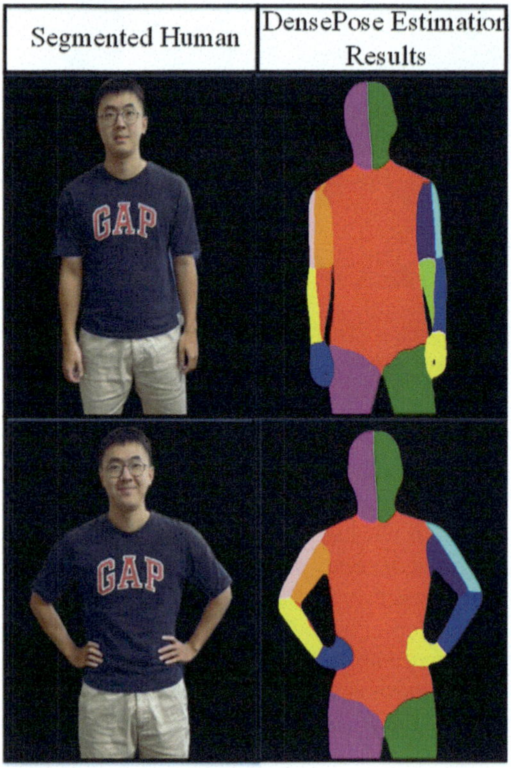

Fig. 5.10 DensePose estimation results on segmented human figure

the Mask2Former model, segmenting the human figure into multiple categories like torso, arms, and legs, as well as the cloth-covered and uncovered areas. The image clearly indicates the level of detail achieved by Mask2Former, paving the way for precise garment deformation in the subsequent stages of our system.

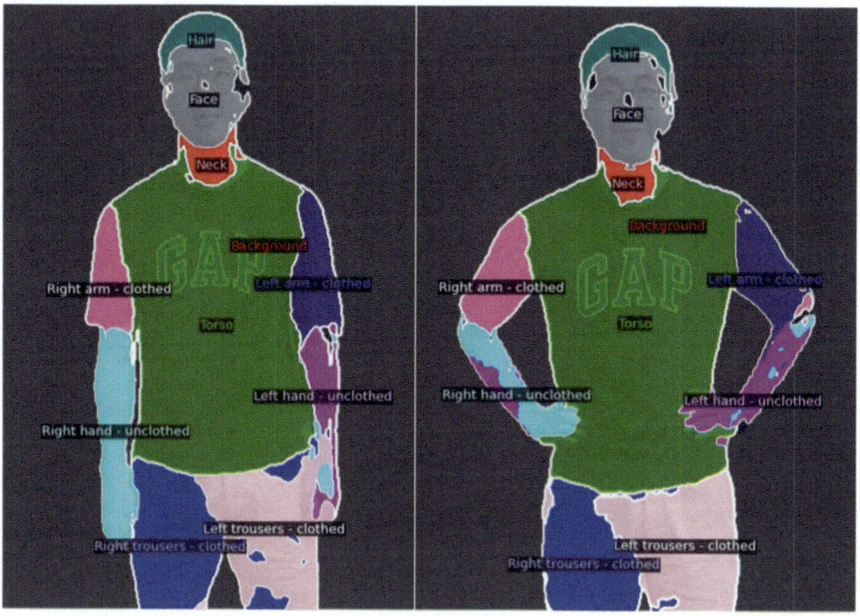

Fig. 5.11 Human and clothing segmentation using Mask2Former

The Mask2Former tool thus proves its utility by offering high-accuracy, fine-grained segmentation that is crucial for individual deformation of clothing during the virtual try-on process.

5.4.5 Virtual Try-on Based on Pre-trained GP-VTON

In this subsection, we highlight the efficacy of using the GP-VTON model for a semantically coherent and highly accurate virtual try-on. Fig. 5.12 shows the results obtained by employing the local-flow global-parsing (LFGP) deformation module in the GP-VTON model for virtual garment fitting.

The segmented human figures from the previous steps serve as inputs for GP-VTON, and the results clearly demonstrate the advantages of using localized flows for independent garment deformation. As can be seen, the LFGP module allows for a highly accurate fit, preserving semantic coherence and avoiding common pitfalls like texture distortion, especially when dealing with complex inputs.

5.4.6 Background Reintegration

The final step in our virtual try-on pipeline is background reintegration, aimed at enhancing the realism of the output by combining the virtually dressed human figure

Fig. 5.12 Virtual try-on results with GP-VTON

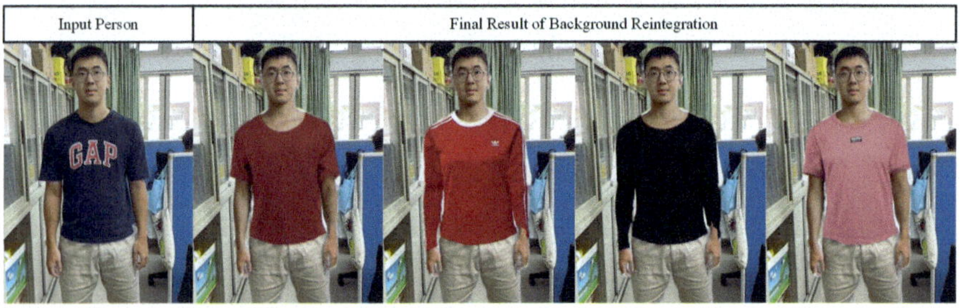

Fig. 5.13 Final result of background reintegration

with its original background. For this purpose, we employ an adaptive inpainting strategy that ensures seamless blending.

Figure 5.13 displays the synthesis result, which is achieved by combining the virtually dressed human figure (obtained from the GP-VTON model) with the originally segmented background. The blend between the dressed human figure and the original background is notably seamless, testifying to the efficacy of our inpainting strategy.

5.5 Conclusion

In this work, we proposed a comprehensive virtual try-on system that leverages state-of-the-art techniques in human segmentation, 2D and 3D pose estimation, semantic segmentation, and background reintegration. Our system effectively tackles the common challenges associated with each stage of the virtual try-on process. We utilized pre-trained models such as the U2-Net for background and human segmentation, OpenPose for 2D pose estimation, DensePose-RCNN for 3D pose estimation, and Mask2Former for fine-grained semantic segmentation. These models were carefully selected to work in tandem, resulting in highly accurate and visually pleasing virtual try-on experiences. Experimental results demonstrated the system's capability to handle a wide variety of human poses and clothing designs, thereby making it applicable for practical use-cases. Additionally, our strategy for background reintegration added an extra layer of realism to the final output. For future work, there is room for further optimization, including the exploration of real-time processing capabilities and the incorporation of additional features such as fabric simulation.

References

1. Qin, X., Zhang, Z., Huang, C., Dehghan, M., Zaiane, O.R., Jagersand, M.: U2- net: going deeper with nested u-structure for salient object detection. Pattern Recogn. **106**, 107404 (2020)
2. Cao, Z., Simon, T., Wei, S.-E., Sheikh, Y.: Realtime multi-person 2d pose estimation using part affinity fields. In: Proceedings of the IEEE Conference on Computer Vision and Pattern Recognition, pp. 7291–7299 (2017)
3. Güler, R.A., Neverova, N., Kokkinos, I.: Densepose: Dense human pose estimation in the wild. In: Proceedings of the IEEE Conference on Computer Vision and Pattern Recognition, pp. 7297–7306 (2018)
4. Cheng, B., Misra, I., Schwing, A.G., Kirillov, A., Girdhar, R.: Masked-attention mask transformer for universal image segmentation. In: Proceedings of the IEEE/CVF Conference on Computer Vision and Pattern Recognition, pp. 1290–1299 (2022)
5. Xie, Z., Huang, Z., Dong, X., Zhao, F., Dong, H., Zhang, X., Zhu, F., Liang, X.: Gp-vton: Towards general purpose virtual try-on via collaborative local- flow global-parsing learning. In: Proceedings of the IEEE/CVF Conference on Computer Vision and Pattern Recognition, pp. 23550–23559 (2023)
6. Han, X., Wu, Z., Wu, Z., Yu, R., Davis, L.S.: Viton: An image-based virtual try-on network. In: Proceedings of the IEEE Conference on Computer Vision and Pattern Recognition, pp. 7543–7552 (2018)
7. Lewis, K.M., Varadharajan, S., Kemelmacher-Shlizerman, I.: Tryongan: body- aware try-on via layered interpolation. ACM Trans. Graphics (TOG). **40**(4), 1–10 (2021)
8. Ge, Y., Song, Y., Zhang, R., Ge, C., Liu, W., Luo, P.: Parser-free virtual try-on via distilling appearance flows. In: Proceedings of the IEEE/CVF Conference on Computer Vision and Pattern Recognition, pp. 8485–8493 (2021)

Chapter 6
Influence of Immersion Teaching on Professional Identity: Taking the Hotel Management Major as an Example

Zhang Lifan and Wang Zhengfen

Abstract In recent years, China has witnessed rapid growth in the number of hotels and the scale of accommodations. However, there exists a shortage of industry professionals, posing a constraint on industry development. Graduates from hotel management programs are also reluctant to choose careers in the hotel industry, leading to a misalignment between industry growth and educational objectives. To bolster students' professional identity, especially at the undergraduate level, certain institutions have embraced an immersion teaching approach in hotel management education, increasing the emphasis on practical instruction. Focusing on undergraduate hotel management students from three colleges in Shanghai, this study employed questionnaires and utilized multinomial logistic regression for data analysis. Focusing on undergraduate hotel management students from three colleges in Shanghai, this study employed questionnaires and utilized multinomial logistic regression for data analysis.

6.1 First Section

In the past two decades, China's service sector has witnessed rapid growth, with a swift expansion in the number of hotels and the scale of accommodations. According to data presented in the Development Report of China's Hotel Industry 2023, released by the China Hospitality Association, as of December 31, 2022, the total number of lodging establishments on the Chinese mainland had reached 430,000, with a staggering 15.291 million guest rooms. Notably, the guest rooms within chain hotels accounted for nearly 5.53 million, marking a robust year-on-year growth of 17.16%. As the hotel industry continues its rapid evolution, it places

Z. Lifan
A Faculty of International Tourism and Management, City University of Macau, Macau, China

W. Zhengfen (✉)
Shanghai Sanda University, Pudong, Shanghai, China
e-mail: wzf@sandau.edu.cn

S.-P. Tseng et al. (eds.), *Advances in Intelligent Information Hiding and Multimedia Signal Processing, Volume 2*, Smart Innovation, Systems and Technologies 416, https://doi.org/10.1007/978-981-97-8760-9_6

increasingly higher demands on both the quantity and caliber of its workforce. However, when considering the supply of talent, job seekers commonly exhibit insufficient enthusiasm for pursuing careers in the hotel industry. This notable scarcity in workforce availability has resulted in an inadequately supplied labor market for the hotel industry, thereby establishing a talent gap that severely hampers, and at times, even restricts the overall progress of the industry. Resolving this intricate challenge necessitates commencing at the root by reinforcing professional identity and elevating occupational quality, thus steering a larger cohort of young individuals toward embracing careers within the realm of hotel management.

In recent years, undergraduate education in hotel management has been undergoing innovative reforms to align with industry advancements. Many institutions are experimenting with immersion teaching, seamlessly integrating practical experiences into theoretical learning. By immersing students in professional scenarios, these approaches enhance students' understanding of potential careers, thereby facilitating their entry into the hotel industry. From the students' standpoint, can immersion teaching truly heighten their sense of professional identity with hotel management? This critical query forms the core of this study. Through distributing questionnaires among students enrolled in hotel management programs, this research examines the impact of immersion teaching on students' affinity for their chosen profession. Ultimately, the goal is to provide insights and recommendations that can enrich hotel management education models, fostering a stronger sense of professional identity among students.

6.2 Literature Review

6.2.1 Professional Identity

Professional identity constitutes a significant aspect of self-identity and personal growth, and it stands as one of the foremost sources of confusion for young individuals. Existing research has examined professional identity from two dimensions. The first is cognitive identity, which refers to "learners' inward acceptance and approval of their chosen field of study based on cognitive understanding." More precisely, it encompasses the emotional reception and approval that learners develop regarding their studies, grounded in their cognitive grasp of the subject matter. This emotional endorsement, coupled with positive external behaviors and a sense of congruity, constitutes a process of emotional, attitudinal, and cognitive integration [1]. The second is behavioral identity, centering on learners' "enthusiastic willingness and proactive engagement in discussions related to their chosen field." It involves demonstrating affirmation and agreement with the chosen profession through active actions [2]. It becomes evident that the development of professional identity involves a dual process. On the one hand, students passively absorb influence and cultivation within the educational environment. On the other hand, they actively acquire specialized ways of thinking through engaged learning, which subsequently transforms into behaviors.

6.2.2 *Immersion Teaching*

Immersion teaching is a pedagogical approach within the field of hotel management education. It positions students at the heart of instruction, fostering a student-centered learning paradigm [3]. It employs a second language as the medium of instruction, simulating future work environments through authentic language, behavior, and scenarios, thereby evoking students' aspirations and anticipation for their future careers [4]. Hotel management education widely embraces the "immersion" mode, leading to the establishment of teaching models that combine academia and industry, typified by institutions like the Université de Lausanne in Switzerland and Cornell University in the United States. This model underscores the close integration of disciplines and industries and highlights the contribution of educational institutions to the hotel sector [5]. The characteristics of the immersive teaching mode are predominantly demonstrated in the following aspects: curriculum design, focusing on cultivating relevant professional skills and management knowledge in the hotel industry; instructional systems, emphasizing the emulation of authentic hotel environments; teaching mode, emphasizing practical components with a significant portion of coursework allocated to practical training, often comprising more than half of total instructional hours [5]; and learning environments, fostering a professional atmosphere to ignite students' interest in learning. Given the rich implications of the immersion teaching, its influence on students' professional identity is multi-faceted. Given the rich implications of the immersion teaching, its influence on students' professional identity is multi-faceted [6]. Additionally, innovative teaching methods like Micro-class and flipped classroom, where students assume the role of active participants, can enhance students' cognitive understanding of their field of study [7]. Other research reveals that on-campus production, internship, and training centers are crucial vehicles for nurturing students' vocational qualities [8]. For hotel management programs that possess a distinct practical nature, on-campus training provides students with an initial exploration of the hotel industry, allowing them to form a tangible perception of the field. Moreover, some studies find that internships during the learning phase notably influence students' future career aspirations [9]. The higher students' satisfaction with their internships, the more inclined they are to choose careers related to their field of study in the future, indicating that higher internship satisfaction correlates with greater professional recognition at the behavioral level. Furthermore, research indicates that the degree of influence of learning interest on learning behavior is substantial, and cognitive comprehension of the chosen field is inevitably influenced by interest [10]. Secondary classrooms and similar approaches more effectively arouse students' interest, facilitating the transition from theory to cognitive understanding [11].

In summary, this study asserts that the immersion teaching can exert a positive influence on students' professional identity. It deepens students' comprehension of the hotel management discipline, bolsters their interest in pursuing careers within the hotel management industry, and puts forth the following hypotheses:

Hypothesis 1–1: Classroom instruction and perceived course content positively influence students' cognitive-level professional identity.
Hypothesis 1–2: Classroom instruction and perceived course content positively influence students' behavioral-level professional identity.
Hypothesis 2–1: The more positive the experiential perception of students in training facilities, the higher their cognitive-level professional identity.
Hypothesis 2–2: The more positive the experiential perception of students in training facilities, the higher their behavioral-level professional identity.
Hypothesis 3–1: The more positive the experiential perception of students in professional internships, the higher their cognitive-level professional identity.
Hypothesis 3–2: The more positive the experiential perception of students in professional internships, the higher their behavioral-level professional identity.
Hypothesis 4–1: The more diverse the comprehensive discipline-related practical activities, the higher students' cognitive-level professional identity.
Hypothesis 4–2: The more diverse the comprehensive discipline-related practical activities, the higher students' behavioral-level professional identity.

6.3 Data Analysis

6.3.1 Survey Sample

Given the progressive nature of hotel management education in Shanghai, along with the adoption of immersion teaching by multiple undergraduate institutions, this study chose current students enrolled in hotel management programs from several undergraduate institutions in Shanghai as the research subjects. A total of 256 valid questionnaires were distributed and collected from three undergraduate institutions offering hotel management programs. Descriptive statistics of the survey sample are presented in Table 6.1.

6.3.2 Variable Measurement

Based on relevant literature and different dimensions of the categorized immersion teaching, four secondary indicators are set to evaluate the degree of implementation of immersion teaching: classroom teaching and curriculum arrangement, on-campus practical training, hotel professional internship, and comprehensive practical activ-

Table 6.1 Descriptive statistics of survey samples

Variables	Percentage	
Gender		
	Male	43.0%
	Female	57.0%
Grade		
	Freshman	11.7%
	Sophomore	23.4%
	Junior	30.5%
	Senior	4.4%
Familiarity with the major		
	Yes	82.4%
	No	17.6%
Involvement of family/friends in the hotel industry		
	Yes	54.3%
	No	45.7%
Desire for major selection		
	Personal preference	48.4%
	Parental preference	22.7%
	Influence of relatives and friends	17.2%
	Willingness to accept adjustment	11.7%
	Sample size	256

ities. The evaluation indicator system is shown in Table 6.2, and a 5-level Likert scale is used for measurement:

The dependent variable of this study is professional identity, categorized into cognitive-level professional identity and behavioral-level professional identity. In the questionnaire, the measurement is carried out through the following questions respectively: "I have increased my identity with hotel management," and "I intend to pursue a career related to hotel management after graduation." These responses are ranked on a scale of 1 to 5, with 1 representing complete disparity and 5 signifying full concurrence.

Furthermore, this study also incorporates gender, academic year, transfer student status, presence of relatives in the hotel industry, and familiarity level with the major as control variables in the regression model. Bartlett's sphericity test and the Kaiser-Meyer-Olkin (KMO) measure were employed for factor analysis of the scales. The analysis results demonstrate that the data sample possesses a high level of structural validity and stability.

6.3.3 Statistical Analysis Results

This study employed a logistic model (multilevel binary logistic model) to analyze the data. Regression analysis was conducted using cognitive-level professional

Table 6.2 Evaluation indicator system for immersion teaching

Primary indicator	Secondary indicators	Indicator content
	Classroom teaching and curriculum arrangement	Mastery of professional terminology
		Proportion of practical courses
		Integration of theory and practice
		Incorporation of case studies in class
		Adequacy of facilities and equipment
		Improvement in professional skills
Immersion teaching mode	On-campus practical training	Utilization of training bases
	Off-campus internship	Intensity of internship work
		Ability to handle internship pressure
		Enhancement of work abilities
		Mastery of work content
	Comprehensive	Frequency of professional competitions
	Practical activities	Frequency of hotel visits
		Frequency of guest lectures
		Degree of enrichment in Club activities

identity and behavioral-level professional identity as dependent variables. The results of the analysis are presented in Table 6.3, with Model 1 utilizing cognitive-level professional identity as the dependent variable, and Model 2 employing behavioral-level professional identity as the dependent variable:

The analysis results from the data model reveal that all four dimensions of the immersion teaching mode exert varying degrees of influence on students' professional identity. Examining cognitive-level professional identity, it's evident that the dimension of classroom teaching and curriculum arrangement plays a significantly positive role in the model. Classroom teaching, through imparting foundational knowledge and its comprehensive role in terms of professional terminology, theory, practical knowledge, and contextual case introductions, plays a fundamental role in shaping students' perceived level of professional identity. This supports Hypothesis 1–1. The perceived experience of on-campus practical training also shows a positive and notable impact. The creation of simulated environments in on-campus training facilities, coupled with skill development activities, aids students in forming initial professional quality and enhancing their professional identity, which supports Hypothesis 2–1. Likewise, campus-wide comprehensive practical activities wield a substantial influence. The relaxing and flexible campus activities serve as vehicles for conceptual immersion, reinforcing students' cognitive-level identity, thus supporting Hypothesis 4–1. However, the perceived quality of professional internships has no significant impact on cognitive-level professional identity, thereby not supporting Hypothesis 3–1. Looking at behavioral-level professional identity, the perceived experience of classroom teaching and

Table 6.3 Regression analysis results of immersion teaching and professional identity

Independent variables	Model 1	Model 2
Perception of classroom teaching course and arrangement	0.327*** (0.089)	0.242* (0.095)
Perception of on-campus training	0.276** (0.084)	0.178 (0.094)
Perception of off-campus internship	0.214 (0.063)	0.354** (0.106)
Perception of comprehensive practice	0.2605** (0.086)	0.278** (0.093)
Control variables		
Gender	0.084 (0.106)	−0.070 (0.115)
Grade	0.119* (0.056)	−0.052 (0.060)
Transfer student status	0.028 (0.190)	0.600** (−0.206)
Familiarity level with the major	−0.321 (0.166)	−0.239 (0.181)
Presence of relatives in the hotel industry	−0.277* (0.107)	−0.512*** (0.115)
Intercept	0.678 (0.580)	0.931 (0.653)
N	256	256
F	15.202	15.964
Df	8	8
P	0.000	0.000

Note: (1) Values in parentheses are standard errors. (2) $^{*}P < 0.05$, $^{**}P < 0.01$, $^{***}P < 0.001$

curriculum arrangement, as well as the perception of comprehensive practical activities, continue to demonstrate a significant positive impact. This underscores the undeniable importance of classroom instruction. The deeper the integration and internalization of knowledge through classroom teaching, coupled with a richer professional learning atmosphere on campus, the higher the level of behavioral-level professional identity, supporting Hypotheses 1–2 and 4–2. Moreover, professional internships also exhibit a significant positive influence on behavioral-level professional identity. This highlights the profound impact of hands-on work experience on individual behavior. Real-world work environments have a stronger potential to stimulate behavioral choices. The more positive the experiential aspect of students' hotel internships, the easier it becomes for them to develop behavioral-level professional identity, supporting Hypothesis 3–2. However, on-campus practical training does not significantly influence behavioral-level professional identity, thus not supporting Hypothesis 2–2.

In summary, the immersion teaching exerts varying degrees of significant influence across multiple dimensions on students' cognitive-level and behavioral-level professional identity. The more effectively the immersion teaching operates, elevating students' sense of acquisition and experience comprehensively, the more it significantly strengthens their sense of identity with the field of study. Furthermore, it motivates them to pursue future career choices aligned with professions related to their chosen field.

6.4 Conclusion

Through analyzing the data on the immersion teaching and professional identity, this study has discovered that the effective implementation of the immersion teaching mode contributes to enhancing students' professional identity. Specifically, the key insights gleaned from this research are as follows: Firstly, classroom instruction and curriculum arrangement exhibit a significantly positive impact on both students' cognitive-level and action-level professional identity. This underscores the continued importance of classroom teaching within the immersion teaching. While the immersion teaching emphasizes the significance of practical instruction, classroom teaching remains a crucial avenue for students to establish and deepen their professional identity. Therefore, reforms and innovations in classroom teaching should continue to be prioritized to provide a strong foundation for practical learning. Secondly, by organizing professional competitions, arranging campus visits, hosting lectures, and facilitating club activities, educational institutions craft a rich professional ambiance on campus. This extends and broadens the process of professional learning, immersing students in their learning environment, and increasing their interest in acquiring professional skills. This approach also contributes significantly to enhancing students' professional identity, making it an indispensable element of the immersion teaching mode. Lastly, on-campus practical training and off-campus internships complement each other, forming the core of practical instruction within the immersion teaching. On-campus practical training helps elevate students' cognitive-level professional identity, while off-campus internships enhance their behavioral-level professional identity. Equal attention should be given to both on-campus practical training and off-campus internships within the teaching process.

Immersion teaching mode has the characteristics of "simulation," analyzing and researching the identity of hotel management students in this mode is more conducive to presenting the role of the teaching mode in shaping the students' psychological identity, and clarifying the psychological mechanism of enhancing students' professional identity for hotel management education. On the one hand, the exploration of the immersion teaching mode not only enriches the theory of hotel management professional construction, but also helps to improve and innovate the existing mode of "school-enterprise cooperation" and explore a better path. On the other hand, it enriches the cross-cutting field of educational psychology and vocational psychology, and provides suggestions for solving the problem of the psychological gap when graduates enter the society.

This study helps to understand the status quo and characteristics of professional identity of hotel management students in colleges and universities, and can prompt colleges and universities to take advantage of existing resources, optimize the use of educational resources, adjust the content and methods of education, and improve the learning environment of hotel management majors. It provides a reference basis for the formulation and adjustment of hotel education policies in colleges and universities. It is conducive to verify the rationality of the teaching mode theory at the empirical level, and can in a certain sense provide suitable case material support for the existing theory, and enhance the theoretical validity of the educational model and the universality of the promotion.

Applied colleges and universities to create a hotel management "immersion" education model, not only in the simulation of real links to improve students' practical ability, but more importantly, through the combination of on-campus training bases and off-campus internships, so that students can obtain a more comprehensive and in-depth understanding of the rationality of the hotel. Cultivate students to really look at the different departments and positions of the hotel from the perspective of "management," and to look at the operation process of all parties from the perspective of hotel senior managers. Enhancing the sense of identity of hotel management students for this specialty is an important contribution to the supply of talents in the hotel industry on the one hand, and is also conducive to improving the establishment of a clear career planning path for students on the other.

It can be concluded that the effective application of immersion teaching contributes to the comprehensive enhancement of students' professional identity. This teaching approach can be widely employed in undergraduate hotel management education to deepen students' holistic comprehension of the profession and industry. It enables them to continuously cultivate and refine their professional skills during the learning process, adequately preparing them mentally and skill-wise for future careers in hotel management.

References

1. Panbo, Q.: Characteristics and Related Research of College Students' Professional Identity. Master's Thesis of Southwest University (2009)
2. Dingming, W., Yongcun, L.: Survey on Master's graduate professional identity. China High. Educ. Res. **8**(12), 18–22 (2007)
3. Yan, Q., Yuyan, S.: Application of immersion teaching in the "management" course. Knowl. Library. **10**(12), 94–95 (2019)
4. Mingzhu, R.: Exploration of "immersion" teaching in cultivating vocational literacy of college students. J. Jiujiang Vocat. Tech. Coll. **2**(12), 89–90 (2016)
5. Xinjun, D.: Comparative analysis and enlightenment of undergraduate education models in Hotel Management in North America and Europe: a case study of Cornell University School of hotel administration and Ecole Hoteliere de Lausanne. J. Sichuan Tour. Coll. **5**(6), 85–88 (2016)
6. Mi, Z.: An exploration of the relationship between vocational cognitive recognition and employment intention of hotel management students in vocational colleges. Forum Ind. Technol. **20**(13), 103–104 (2021)

7. Yunzhong, X.: Analysis of professional identity and influencing factors of sociology graduates. J. Chengdu Univ. Technol. (Soc. Sci.). **19**(03), 93–99 (2011)
8. Rong, G.: Exploration of the construction model of in-school productive internship training bases in vocational colleges. For. Educ. China. **29**(01), 43–45 (2011)
9. Yi, Z., Binbin, K.: The impact of student on-the-job internships on career intentions: the mediating effect of psychological contract. Enterprise Sci. Technol. Develop. **2**(04), 205–208 (2017)
10. Dan, W., Chang, L.: Relationship between college Students' learning motivation and learning behavior: an empirical study in the field of ideological and political education in universities. J. Southwest Univ (Sci. Ed.). **43**(03), 88–94 (2017)
11. Huijuan, L., Chaoyi, C.: Construction of the secondary class activity system in Hotel Management in Vocational Colleges from the perspective of quality iceberg. Campus. **11**(29), 107–109 (2018)

Chapter 7
Design and Implementation of Hotel Management System Based on Uniapp Framework

Tong Jun, Liuqiong Ping, and Zaiying Liu

Abstract In order to improve the efficiency of hotel management business, a multi-terminal and controllable hotel management system has been designed and implemented in a targeted manner. The system can meet the current and future requirements for efficient collection, storage, processing, organization, management, and utilization of information resources, achieving a more efficient and convenient management model. The PC front-end uses Vue and Element frameworks, the mobile front-end uses Uniapp framework combined with WeChat mini program technology, and the back-end uses Spring Boot framework and integrates multiple open-source technologies. The system has been applied to multiple chain hotels, running stably and reliably, and has strong practical value.

7.1 Introduction

Design and implement a management system for chain hotels with a wide range of complex business types in the hotel industry, to address the shortcomings of traditional manual processing methods used by hotel management personnel. The system can provide accurate and timely information on check-in tenants, guest room information, guest room cleaning status information, and report statistics displaying various types of data for management personnel; At the same time, it provides an efficient and convenient electronic means for information exchange and daily hotel management among personnel at all levels of the hotel; The ultimate goal is to further improve the cooperation level of hotel personnel at all levels, enhance the management level and office efficiency of department managers, reduce work burden, meet the current and future needs for efficient collection, storage,

T. Jun (✉)
Shanghai Sanda University, Shanghai, China

L. Ping · Z. Liu
School of General Courses Shanghai Civil Aviation College, Shanghai, China
e-mail: pingliuqiong@shcac.edu.cn

S.-P. Tseng et al. (eds.), *Advances in Intelligent Information Hiding and Multimedia Signal Processing, Volume 2*, Smart Innovation, Systems and Technologies 416, https://doi.org/10.1007/978-981-97-8760-9_7

processing, organization, management, and utilization of information resources, and achieve a more efficient and convenient management mode.

7.2 System Requirements Analysis

7.2.1 System Functional Requirements

Functional requirements are the core part of software requirements [1]. The functions of a hotel management system mainly include basic data module, tenant management module, business management module, revenue management module, expenditure management module, data statistics module, and file management module, as shown in Fig. 7.1.

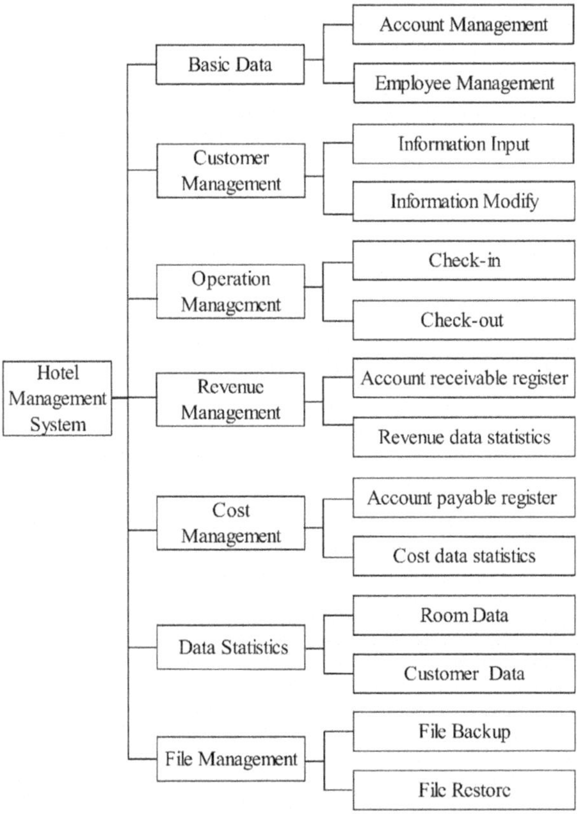

Fig. 7.1 System function modules

7.2.2 Non-functional Requirements

To ensure long-term, safe, stable, reliable, and efficient operation, the hotel management system should meet the following performance requirements:

Usability The system's page is user-friendly, fast, effective, and easy to maintain, with data statistics and chart display functions [2].

Security System deployment security, data backup, and disaster recovery, using login authentication and core data encryption strategies [3].

7.2.3 System Role and Process Analysis

The hotel management system involves four types of users: super administrators, administrators, staff, and regular visitors. The system adopts a design approach where some user permissions are compatible downwards and some users have specific permissions. For example, a super administrator has the highest permissions and the interface can display all functional modules. The interfaces of staff and ordinary users will only display their respective functional modules.

Meanwhile, as the system is specifically designed for use by hotel internal management personnel, there is only one super administrator user and it is independently initialized from the database. The remaining users are created by creating permissions for backward compatible users, while higher-level users can create lower-level users, assign lower-level user permissions, and control the account status of lower-level users. The business process overview of the system is shown in Fig. 7.2.

7.2.4 Technical Analysis

The system conducts technical analysis from various aspects such as PC front-end framework, Mobile front-end framework, back-end framework, and database.

PC Front-End Framework Vue is a lightweight framework that can perform bidirectional data binding and has functionality for code encapsulation and reuse [4]. Combined with document configuration, Vue can be used out of the box to achieve rapid prototyping of the front-end interface.

Mobile Front-End Framework Uniapp can publish a set of code to multiple platforms such as IOS, Android, Web, and various mini programs [5, 6]. The Mobile front-end is developed and compiled based on the Uniapp and UView frameworks for users to use.

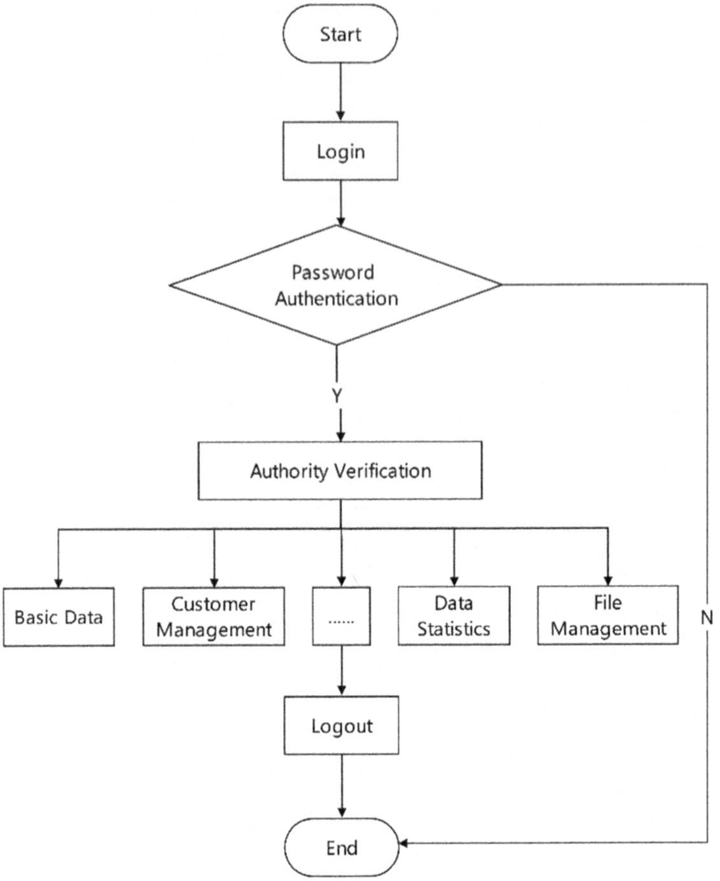

Fig. 7.2 Overview of system business process

Back-End Framework The back-end is developed using the Spring Boot 2.0 framework, which can quickly parse and integrate, use start dependency and automatic assembly mechanisms, and embed tomcat, simplifying a lot of repetitive work outside of non-business logic development [7].

Database Using MySQL database and Redis can achieve data storage and persistence. MyBatis Plus supports custom SQL, stored procedures, and advanced mapping, which can effectively improve development speed [8].

7.3 System Design

7.3.1 System Architecture Design

The system adopts front-end and back-end separation technology. The back-end uses the Spring Boot framework, and the back-end's permission authentication uses the Shiro framework to distribute JWT authorization; the PC front-end adopts Vue and Element frameworks, while the mobile front-end adopts Uniapp framework combined with WeChat mini program technology. The system is designed from top to bottom as the presentation layer, communication layer, service layer, persistence layer, and data layer. The functions of each layer are different, reducing the coupling of the system, as shown in Fig. 7.3.

Users can request the system front-end through the PC program or mobile App, and log in and authorize through JWT + Shiro, allowing users to perform different operations on different businesses. The data generated by the business will be saved in the MySQL database and MinIO file storage system on the cloud server.

The reasons for choosing the architecture include:

1. The overall structure will be clearer.
2. The separation architecture can significantly improve adaptability for multi-end access.
3. Both front-end and back-end can process data to improve response speed.
4. The front-end is the same as the template, but the asynchronous loading method accelerates data rendering and improves user experience [9].

7.3.2 System Security Design

For the entire system, security is crucial. The system supports secure authentication of user identity, encryption of user private information, control of remote database access, and restriction of database operation permissions, ensuring system security from multiple aspects.

The system integrates Shiro as the security framework. When unauthorized visitors maliciously access the no permission page, the warning of no permission will pop up and redirect to the login page. At the same time, it can also prevent unauthorized access between users with different permissions. After successful login, users will receive a unique identification token distributed in the background and stored in the local browser. JWT consists of three parts: header, payload, and signature. The JWT signature part of the system is encrypted by a unique and confidential Secret string, and each token is set with a validity period of 12 hours. If the token is invalid, a new login is required. Each user's JWT will be stored in Redis to ensure that an account can only log in to one location at the same time [9].

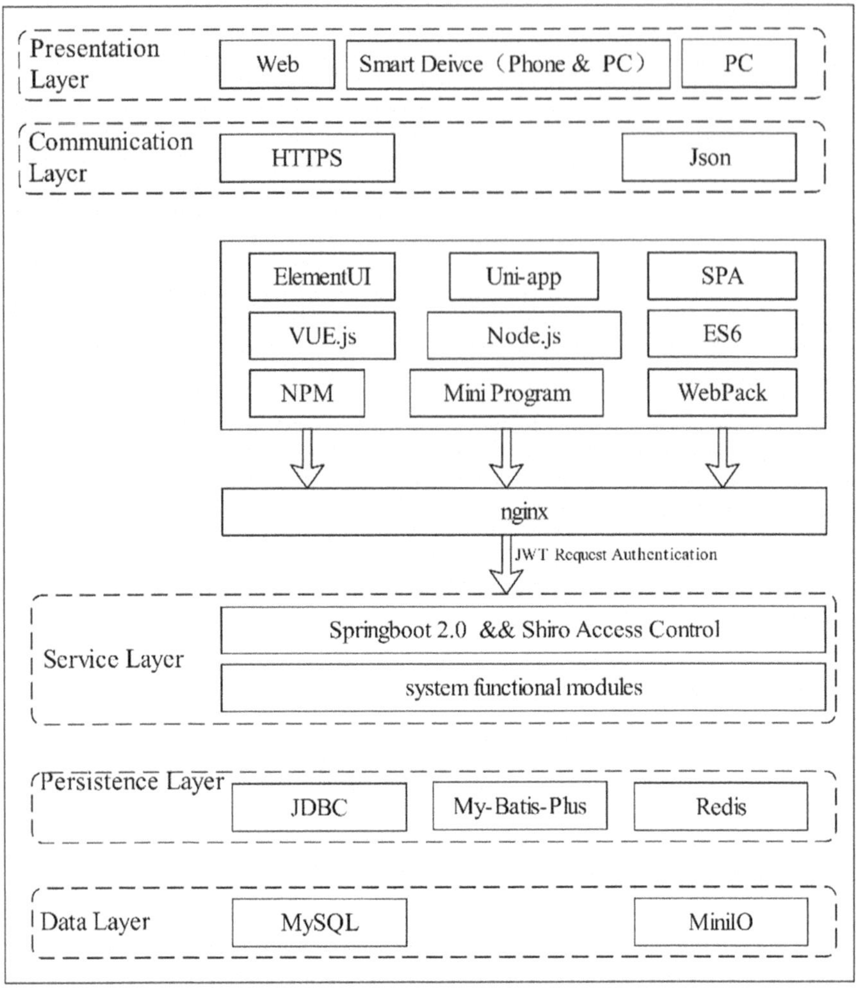

Fig. 7.3 System architecture diagram

7.3.3 Database Design and Interface Design

The system uses MySQL database to store business data, and a large number of backend interfaces directly transmit JAVA class objects for front-end and back-end interaction, with class objects and table objects corresponding to each other. When designing specific table structures, we will strive to meet high paradigms and also consider the operational efficiency of operating the database when the subsequent data volume increases.

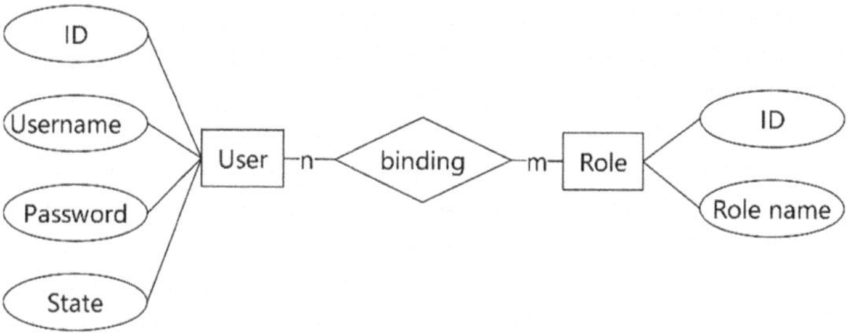

Fig. 7.4 User permission E-R diagram

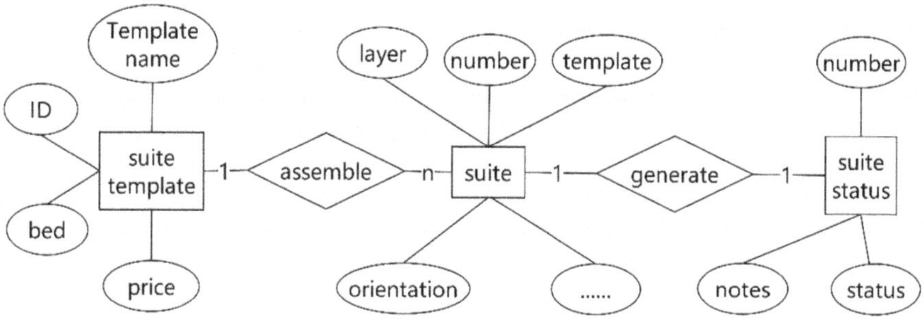

Fig. 7.5 Suite E-R diagram

User Table and Permission Table Design The interface for user information operations will encapsulate data into user classes for transmission. The user permission E-R diagram is shown in Fig. 7.4.

Suite Interface Design The system will encapsulate the data queried by SQL statements into a RoomInfo class. The suite template table can store all types of existing suites, making the information of specific suites more concise, effectively reducing data redundancy, increasing template scalability, and reducing server storage pressure. Due to the administrator's extensive modification and query of suite status, an additional suite status table is created here for business operations, which can effectively reduce retrieval and suite modification time. The suite E-R diagram is shown in Fig. 7.5.

Tenant Information Table and Check-in Registration Table Design If the tenant import function is used, the pre-check-in record for the corresponding date of the tenant can be automatically generated. The personnel information-related fields appear in both tables. The design consideration is that when importing the tenant information table, a check-in registration form will be generated, but a large number of businesses are based on check-in registration form operations, such as check-in, check-out, sorting, export, etc. If the tenant ID is used as a logical foreign key to bind two tables, business operation speed will be affected after a large amount of data is entered in the future. So, all the information of the tenant will also be saved in the registration form as a complete business table.

The design of other business functions is similar to the above and will not be elaborated here.

7.4 Implementation of Key System Technologies

7.4.1 Implementation of User Login Authorization

When a user logs in for authorization, the system back-end will verify: first, verify the existence of the username, and then verify whether the password is correct. After the user successfully logs in, they will first store the JWT in Redis, then put the JWT in the response header and return it to the user end, and save it in the browser's local Storage through front-end code.

The generation and parsing of JWT will be completed by methods in JwtUtils, and the JWT configuration will be extracted into the overall configuration file of Spring Boot. The system will place the JWT returned after successful login in the response header as a key value pair. Because the second payload of JWT can be decoded by base64, it only stores the user's ID and role, and does not return private information such as the user's password. The third signature of JWT will use the secret encryption in the configuration and set the expiration validity time to ensure safe and reliable information transmission.

If a token is carried in the request header, Shiro security manager will parse the JWT to determine whether the token is within its validity period, and authenticate and authorize it based on the permission data carried in the second payload. Then, the request will continue to access the interface, and the users will be judged whether they have sufficient permissions by annotating the required permissions on the interface class. It should be noted that although JWT can alleviate the pressure on the server to store user information, it always carries a local token in the request header, and Shrio security manager will perform authentication and authorization operations on the user every time the interface is accessed.

7.4.2 Integration of Spring Boot and Mybatis Plus Frameworks

The integration of Spring Boot and Mybatis Plus frameworks is involved in all back-end businesses, simplifying the tedious steps of traditional JDBC persistence layer operations. The integration steps are as follows: Firstly, it is necessary to import the.

Mybatis Plus dependency in the POM file, then add a configuration class for the paging plugin, and finally configure the mapper package path in application.yml to start using it. Due to its powerful functionality, during development, any operation on the MySQL database can only be done through annotations, such as adding @Select annotations to the mapper for query operations. This saves developers time writing a large amount of XML configurations by using the underlying reflection mechanism through MP. Mybatis Plus provides all basic interfaces for developers to directly call, and can automatically generate code generator operations for entity classes through the configuration database, greatly improving development efficiency.

7.4.3 Implementation of Suite Module

Due to different businesses, there will be a large number of inquiries related to suite information, availability status, and other related suite operations. In order to avoid poor user interface experience or system crashes caused by a large amount of data redundancy, and to ensure the universality and scalability of the suite module in the later stage, the system adopts the suite module to implement corresponding functions.

The system creates different suite templates and assigns matching suite templates to each suite. At the same time, a corresponding suite status table record and a checkin intermediate table record will be generated for each suite establishment, which are respectively used for the suite status overview interface and suite check-in judgment. This can not only reduce data redundancy, save storage space on database servers, but also improve the efficiency of query operations when data increases in the future.

7.4.4 Implementation of Tenant Management Module

According to actual business requirements, the system first needs to display the information of all tenants in a clear and orderly manner in the functional interface, and then can manually add and customize the function of editing existing tenant information. Due to the need for chain hotels to interface with other departments and obtain long-term tenant information and their check-in dates in advance, the system provides a user-friendly function of importing tenant information based on a fixed template Excel document. Therefore, the system will analyze the tenant information and all scheduled check-in dates and times during the import process into Excel, and organize the data into the tenant object information and each corresponding check-in registration record (which can be used for other businesses) for storage in the database.

We mainly use Spring Boot to integrate EasyExcel technology. Firstly, create the Tenant class based on the existing Excel template and add annotations on the member properties. Through this method, a connection can be established between the system and the Excel file. The system will read the data of each row based on the configured header information and encapsulate it into the Tenant class. Secondly, based on the encapsulated Tenant class, parse the corresponding fields, cut the string to obtain all check-in times, and place it in Set. Finally, perform functional optimization. If the same file is imported repeatedly, duplicate information will not be added. At the same time, the system also supports multiple sheet imports.

7.4.5 Implementation of Front-End Framework

The front-end framework utilizes the Vue component reuse feature and Vue Router mapping to implement single-page applications. Add a fixed left navigation bar and top navigation bar on a single page, and then switch between different functional interfaces through the configured Vue Router. The data request adopts Axios asynchronous request, and the system stores permission data in Vue.

The implementation of the computer functional interface adopts the Element-UI and ECharts frameworks. The mobile interface adopts Uniapp and UView UI frameworks. Basically, the design and implementation of the mobile end are the same as those of the computer end, relying on Vue's component reuse. The design of the navigation bar will be displayed differently on both ends based on the role of the logged-in user. Uniapp is a framework that uses Vue to develop all front-end applications. It can publish a set of code to multiple application platforms such as IOS, Android, Web, and various mini programs [5–6]. The UView UI is a fully compatible Uniapp ecosystem framework with Vue, making the development process convenient and fast.

7.4.6 Integrated Deployment of the System

The system adopts the integrated mode of simple DevOps operation and maintenance development. By continuously integrating the tool Jenkins as the main role in the DevOps loop, CI/CD objectives are achieved. During the development process, new features or important system versions will be uniformly uploaded to Gitee for version management. At the same time, Jenkins deployed on a Linux cloud server can be run to parse the configured Jenkins file to pull the front-end and back-end code saved in Gitee, and then the project can be built using the build tool configured in Jenkins. Finally, deploy and publish the system project through the sh code in Pipline. Docker and Docker Compose containerization technologies have simplified the deployment of this system in the Linux environment. For example, MySQL, MinIO, and Nginx are all managed through Docker Compose technology for unified deployment.

7.5 Conclusion

Starting from the actual needs of the hotel industry business, the article adopts the J2EE development platform and MySQL database for development. The design of the hotel management system was implemented through the Spring Boot and Uniapp frameworks, targeting the needs of super administrators, administrators, staff, and ordinary visitors. The basic data module, tenant management module, business management module, income management module, expense management module, data statistics module, and file management module were designed. The test results show that the first response time using an external network and the first response time using multiple users are both less than 20 ms, and the system can operate stably, safely, and reliably, which is worth promoting and applying. At the same time, we will consider introducing more advanced workflow technologies in the later stage, such as using online big data analysis and mining techniques to analyze workflow logs and improve system efficiency, in order to better complete hotel collaborative business.

References

1. Zhong, N., Yao, Y., Kakenoto, Y.: Automatic construction of ontology from text databases [J]. Data Mining. **2**, 173–180 (2011)
2. Elisa, N., Yang, L.Z., Chao, F.: A framework of blockchain-based secure and privacy- preserving E-government system [J]. Wirel. Netw. **7**, 1–11 (2018)
3. Kim, H.Y., Cho, J.S.: Data governance framework for big data implementation with a Case of Korea [C]. In: 2017 IEEE International Congress on Big Data (Bigdata Congress), pp. 384–391. IEEE, Honolulu (2017)
4. Chandwadkar, D.M., Ashok, C.S., Ugale, S.P.: Industrial parameters monitoring system based on embedded web server [J]. Int. J. Innov. Technol. Exploring Eng. **8**(9), 2450–2453 (2019)
5. Pimentel, V., Nickerson, B.G.: Communicating and displaying real-time data with WebSocket [J]. IEEE Internet Comput. **16**(4), 45–53 (2012)
6. Li, D., Mei, H., Shen, Y.E.: Charts: a declarative framework for rapid construction of web-based visualization [J]. Visual Informatics. **2**(2), 136–146 (2018)
7. Wei, M., Yang, X., Mao, J.: Secure framework and security mechanism for edge nodes in industrial internet[C]. In: Proceedings of the 13th International Conference on Ubiquitous, pp. 254–266. Springer, Cham (2019)
8. Song, F., Zhu, M., Zhou, Y.: Smart collaborative tracking for ubiquitous power IoT in edge-cloud interplay domain[J]. IEEE Internet Things J. **7**(7), 6046–6055 (2019)
9. Matsui, T., Yamamoto, K., Sumi, S.: Detection of lightning damage on wind turbine blades using the SCADA system [J]. IEEE Trans. Power Del. **36**(2), 777–784 (2021)

Chapter 8
Personalized Learning Support Based on Learning Behavior Big Data Analysis in Teaching Resource Library

Xiao-yan Tang, Cheng-jiang Zhang, Bin Liu, and Qing Ye

Abstract Personalized learning respects different education and promotes the all-round development of students, which is the essence of education. Learner data accumulated in the teaching resource libraries implies some goals, interests, needs, and expectations. Guided by cognitive theory and using big data technology, a statistical analysis of learners' behavior characteristics is presented, learners' possible goals and preferences are summarized to predict learning interests and behaviors, and provide learning strategies and learning resources, which provide decision support for the construction of individualized intelligent learning system.

8.1 Introduction

Teaching resource library is the core support of high-quality network teaching, which is the key to improve the level of network teaching and directly guides the in-depth development and future trend of network teaching. The Chinese national vocational education professional teaching resource library is a key educational informatization project supported by the "Special Project for Professional Capacity Building of Vocational Education" of Chinese Ministry of Education, and also is an important means to promote the comprehensive application of information technology in the professional teaching field of vocational education [1].

Personalized learning refers to the learning paradigm based on reflecting students' personality differences and aiming at promoting students' personality development [2]. Personalized learning embodies fairness respects the difference of education and promotes the all-round development of students, which is the essence of education. Due to the large differences in many factors such as learner's identity, experience, learning basis, learning objectives, and learning conditions, the personalized learning system which can provide guidance and convenience

X.-y. Tang (✉) · C.-j. Zhang · B. Liu · Q. Ye
Changzhou College of Information Technology, Changzhou, China

S.-P. Tseng et al. (eds.), *Advances in Intelligent Information Hiding and Multimedia Signal Processing, Volume 2*, Smart Innovation, Systems and Technologies 416, https://doi.org/10.1007/978-981-97-8760-9_8

for learners is paid more and more attention. With the normalization of network learning, a large amount of learning behavior data, including temporary stream data and long-term static data, has been formed in the resource library. As a learning data set, big data analysis technology can be used to extract, process, and summarize the characteristic data of learning behaviors to form a classified knowledge database.

In Sect. 8.2, we discuss related work. In Sect. 8.3, a support and improvement of personalized learning by big data analysis is presented. In Sect. 8.4, we propose a personalized learning data architecture based on big data analysis and decision support. In Sect. 8.5, we discuss the personalized learning support practice under decision support and make an experimental analysis. In the last section, we conclude our paper and discuss our planning for future work.

8.2 Related Work

In recent years, scholars have done a lot of research and practice around personalized online learning, especially the study of e-learning behavior, including the standardization, analysis, and modeling of e-learning behavior data [3], mining of online learning behavior data, etc. [4]. E-learning behavior is a variety of learning-related behaviors that occur in e-learning environment, involving psychology and behavioral science, pedagogy, and information science technology [3]. put forward eight behavioral elements, adding three elements of "behavioral purpose, behavioral process, and behavioral intensity" based on the five basic elements of "behavioral subject, behavioral object, behavioral environment, behavioral means, and behavioral result" [4]. put forward the view that network learning behavior is divided into three categories: information retrieval processing behavior, problem-solving behavior, and interpersonal communication behavior. But there are still many problems that need further research and practice.

The standardization and effective extraction of online learning behavior data need to be further improved. The correlation between network learning behavior and learning effect lacks theoretical basis, and it needs comprehensive analysis of multi-disciplinary research and verification of actual network teaching data. The communication between learners and teachers is not enough, and the pertinence of guidance is not strong enough. The degree of machine intelligence is not deep. The large amount of data in the existing teaching resource base has not been effectively analyzed and utilized. The reduction and generalization of data also need the guidance of psychological and behavioral theories. The particularity of professional teaching network in higher vocational education, especially the related problems of skill training, needs to be further studied.

8.3 Support and Improvement of Personalized Learning by Big Data Analysis

Guided by constructivism, in accordance with the principle of “teacher-led and student-subject”, and based on American psychologist Bandura’s social cognitive theory of “interaction among behavior, environment, and subject factors” [5], the overall characteristics of learning behavior are comprehensively analyzed to discover the interest points, learning preferences, professional level, learning speed, goal orientation, etc. that may be associated with individual learning behaviors, and then to construct an intelligent engine which can provide heuristic guidance, give learning modes suggestions, and customize learning resources for learners, etc. to establish an intelligent interaction mechanism and provide basic conditions for personalized learning decision support.

8.3.1 Features of Learning Behavior Big Data

The teaching resource library has a clear business logic flow and data flow which can implement a complete record of learners’ access process. The system can obtain individual learning behavior data including: learner ID, login time, learning duration, learning resource ID, learning resource type, exercise selection, test or assessment results, class speech, and suggestions. The learning behavior data can be divided into structured data, semi-structured data, and unstructured data according to the storage mode, and also can be divided into streaming data and static data according to the time characteristics.

Learning behavior data can, to some extent, reflect learners’ habitual learning period, learning progress, learning preferences, expecting goals, focus points, relevant knowledge, problem-solving ability, application ability, understanding depth, and other characteristics. Both the complete set or subset of learning behavior data can be selected and summarized according to certain strategies to form feature reference values, which can be stored as long-term data and used to build a knowledge base. The individual data can be dynamically extracted and analyzed, so as to summarize the reference values of learners’ personalized learning.

8.3.2 Value of Learning Behavior Big Data

In the construction and use of teaching resource library, educators and system developers have been exploring ways to improve the quality of online teaching, committed to building a high-quality intelligent learning platform. They have accumulated a lot of experience and gradually deepened the understanding of the internal relationship between online learning behavior and learning effect, and made a lot of attempts in improving teacher-student communication, strengthening collaborative learning and exploration learning.

In recent years, the rapid development of database, machine learning, and big data analysis and visualization technology has provided strong support for the collection, cleaning, transformation, screening, storage, labeling, quantitative and qualitative analysis, and multidimensional display of network learning behavior data. According to the results of big data analysis, the experience of teachers, and the opinions of learners, a data model of "learning behavior characteristics–learning guidance" can be established, and an intelligent interactive mechanism can be constructed to improve the autonomous learning function of teaching resource library. Based on empirical probability data, this mechanism can analyze learners' styles and predict learners' behaviors [3], so as to provide customized learning support, recommend learning modes and resources according to learners' personalities and preferences, and optimize learning paths. Therefore, the decision support function of the teaching resource base can be gradually expanded and strengthened, and the personalized learning system is improved and enriched. Under the guidance of constructivism theory, the analysis of behavioral characteristics of network learning based on big data has played a far-reaching revolutionary role in building an intelligent, harmonious, high-quality, and efficient network teaching system that supports personalized learning.

8.4 Personalized Learning Data Architecture Based on Big Data Analysis and Decision Support

The personalized learning data analysis and decision support system includes analyzing the overall characteristics and trends of the learning behavior, comparing and finding the correlation pattern among the habit characteristics, interest trend and learning goal of individual learning behavior, and establishing an intelligent individualized learning support model.

Based on abundant online learning resources, the experimental method and big data technology are used to analyze learners' learning style, knowledge focus, and habitual learning time, so as to formulate learning guidance strategies and determine recommended resources and push time.

8.4.1 Learning Behavior Association Description and Data Model

Based on the comprehensive analysis of the current construction and operation of the teaching resource library, part of the relevant learning behavior data is extracted as the characteristic data, and the analysis and decision data model is built based

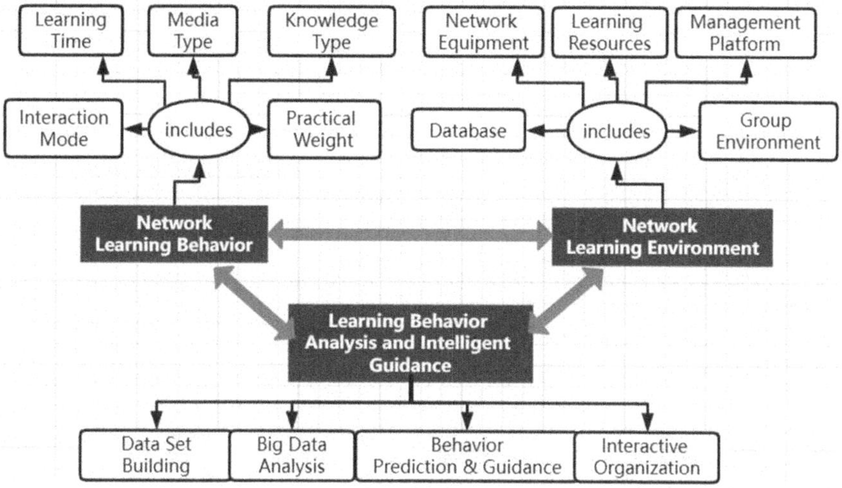

Fig. 8.1 Conceptual model of learning behavior characteristics–learning guidance

on cognition and behavior theory. Based on Bandura's social cognitive theory of "behavior, environment and subject factors influencing each other [5]", to predict learning goals and learning preferences according to the characteristics of learning behaviors, explore how to improve the online learning environment, focusing on improving interactive strategies and methods and intelligent recommendation of learning resources, so as to provide convenient and effective guidance and help for learners in the context of rapid increase and great abundance of online learning resources, and affect learning psychology and learning behavior to improve learners' autonomous learning efficiency. Therefore, the conceptual model of "learning behavior characteristics–learning guidance" is constructed, as shown in Fig. 8.1.

Due to the multi-faceted nature of network learning behavior data and the diversity and dispersion of record storage, the data is pre-processed to extract some components with typical characteristics, easy to understand the semantics and easy to associate with the learning effect. According to the principle and technology of data warehouse, data cubes in a multidimensional form are combined. The purpose is to facilitate multi-angle and multi-level comprehensive analysis of learners' learning objectives, learning preferences, learning priorities, learning states, learning effects, etc., and provide decision-making basis for suggesting learning programs and learning resources to support personalized learning.

After pre-processing such as extraction and conversion marks, the characteristic data of learning behaviors are organized in a multi-dimensional way, including learner ID, gender, access time, learning resource ID, knowledge type, media type, practical operation degree, independence degree, etc. The basic data unit is learning duration. Learning resources are classified according to their various characteristics. The relevant data containing the evaluation of learning results is composed of a

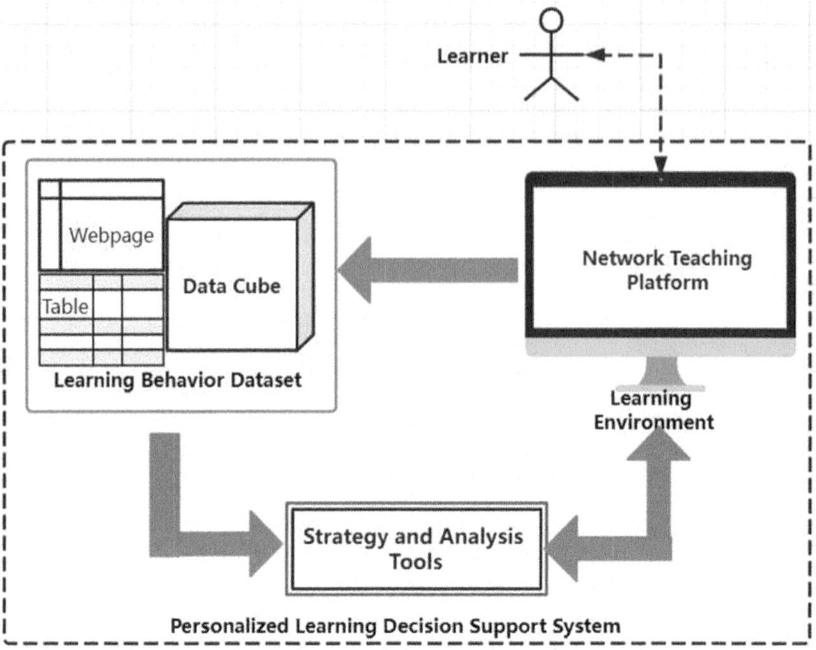

Fig. 8.2 Data model of learning behavior characteristics–learning guidance

two-dimensional table, including learner ID, learning resource ID, access time, and grade, as shown in Fig. 8.2.

8.4.2 Data Flow with Personalized Learning Path

The original data of learning behavior are pre-processed by data cleaning, correlation analysis, labeling, data transformation and convention, generalization, etc., and two-dimensional tables and data warehouses are constructed to form the training data set of learning behavior. After comparative analysis, classification and clustering algorithms are selected and applied to construct the classifier, and process the learning behavior training data set. The classifier is evaluated and modified with a randomly obtained test data set.

Based on the practical teaching experience of teachers, and questionnaire survey results of learners, and the statistical analysis results of specific characteristics of online learning behaviors, relevant concepts, and quantitative thresholds are stored to form a knowledge base, which can be used as the basis for the classification of individual learning behaviors. The personalized learning resource recommendation strategy is designed and the support engine is constructed. The engine can be automatically triggered to test the individual learning behavior data by using the classifier and get the category characteristics of learners. According to the classification results, learning schemes and learning resources are regularly recommended to learners as reference.

8.5 Personalized Learning Support Practice Under Decision Support

8.5.1 Learn Feature Discovery

Based on the historical data in the teaching process of "Higher Vocational Education Software Technology Professional Teaching Resource Library", this paper adopts the method of combining description and prediction, using the method of illustration, comparison, and experiment, and firstly finds out the group commonality of learners in different data abstraction layers and classifies them. Taking this as a reference, the characteristics of individual learners are compared and classified to predict learning behaviors and provide personalized support.

The following basic information about learners and behaviors was obtained from multiple dimensions. There were 7648 learners in total, and the learning resources were counted according to knowledge categories: 123,246 in theory and 165,511 in skill.

Learning preferences were simplified as access time. Quantiles were set in descending order by access time (seconds) or test score (percentage system), and preference thresholds were set in combination with teacher experience (Q1 was strong, Q3 was weak) (see Table 8.1). According to this classification method, the classification of learning behavior features is formed and the classifier is constructed, as shown in Table 8.1.

8.5.2 Evaluation Model and Individual Learning Behavior Data Analysis

According to the results of data analysis, six dimensions of typical learning behavior characteristics are selected to construct an evaluation model, and the specification processing is done based on the original data. According to the evaluation model, a classifier is used to conduct a comprehensive analysis of individual learning behaviors, and learning preferences can be described more vividly, as shown in Table 8.2.

Various strategies can be adopted to recommend learning programs and learning resources to learners, or explain the limitations in the use of learning resources, or reflect the problems such as insufficient resource construction. In this example, the

Table 8.1 Clustering of learning behavior feature data

	Firs quantile (Q1)	Median (Q2)	Third quantile (Q3)	Strong threshold (Q1)	Weak threshold (Q3)
Text (sec.)	173	28	7	173	7
Graphic/image (sec.)	184	34	8	184	8
Audio (sec.)	558	162	48	558	48
Video (sec.)	796	277	91	796	91
Animation (sec.)	162	34	12	162	12
Virtual simulation (sec.)	63	21	2	63	2
PPT (sec.)	227	68	39	227	39
Other material (sec.)	552	239	80	552	80
Exercise (score)	77.3	62.8	26	77.3	26
Test (score)	87.8	74.3	35.3	87.8	35.3
Discussion (each)	79	4	1	79	1

Table 8.2 Individual learning behavior characteristics (access duration (seconds))

Learner ID	Text	Video	Animation	PPT	Other	Test
ID117	284	1195	117	215	592	93.4
ID195	74	359	15	185	320	76.7
ID949	31	23	0	48	105	47. 1

Table 8.3 Learners focus on resources in a certain period of time

Learner ID	Resource ID	Major ID	Course ID	Resource Name	Concern Degree	Skill Weight	Access Time
ID117	404,509	1247	13	Unit 5 construct type	1	1	11:00
ID195	1927	29	5	Programming_ video Production of tunnel design	1	1	13:00
ID949	406,584	1247	13	For felt Mo Shan- video Lecture 18 Expression_PPT	0.6	0.6	11:00
ID949	406,585	1247	13	Lecture 19 Flowchart_PPT	0.3	0.6	11:00

three learners embody relatively different characteristics, which provides specific reference for teachers to improve the teaching process and guide learning. The classifier is used to determine learners' personal preferences, locate specific learning resources, extract relevant information, and comprehensively describe learners' characteristics, so as to serve as the recommendation basis for learning resources, as shown in Table 8.3.

8.5.3 Personalized Learning Resource Recommendation

According to the characteristics of individual learning behaviors, relevant operations are carried out to select and recommend learning resources, as shown in Table 8.4.

Table 8.4 Recommended learning resources

Learner ID	Latest learned resource ID	Resource type	Preference	Skill weight	Recommendation	Access time
ID117	404,509	Video	1	0.9	404,513	11:00
ID195	1927	Video	0.9	1	1939	13:00
ID949	406,584	PPT	0.6	0.7	406,594	11:00
					406,595	

8.5.4 Experimental Analysis

The simplified and typical analysis of learning behavior characteristics in the teaching resource library is beneficial for effective data processing. Part of the learning behavior data of the teaching resource library is extracted for pre-processing, specification, and generalization to form a training data set that can be used for analysis and mining, and the classifier is constructed and the learning behavior is predicted. The results of the experiment partially verify the preconceived idea and basically achieve the purpose of the experiment.

In the questionnaire and interview survey of this experiment, 50.64% of the participants strongly agree with personalized learning resource recommendation based on learners' preferences.

8.6 Conclusion

With the rapid development of information technology, learners' expectation on the quality of network teaching resources is also increasing. Online teaching is playing an increasingly important role and has become an indispensable way and means for learners [4]. In the context of non-face-to-face teaching, the interaction between learners and teachers needs to be improved and the interaction content needs to be enriched. The instruction of teachers is faced with many problems, if we can provide dynamic and targeted intelligent support, it will have a great effect on improving the quality of teaching.

At present, it is difficult to record and express the psychological characteristics of learners in e-learning behavior data, and it is difficult to realize the formal understanding of many learning behaviors, and a large number of data analysis experiments are needed to explore and verify their exact meanings. It is particularly important to deeply understand the correlation between e-learning behavior data, learning goals, and learning effects. The analysis and mining techniques for e-learning behavior data need to be constantly improved and tested in practice. In the future, a variety of technologies will be used for comprehensive analysis and comparison to obtain more valuable information to provide self-learning intelligent decision support more effectively.

Acknowledgments This work is partially supported by Jiangsu University "Blue Project", Jiangsu Vocational Education Research Project "Research on the Application of Big Data Analysis Technology in Improving Learning Effect of Teaching Resource Library" (XHYBLX2023021), Jiangsu Vocational Education Research Key project "Research on the Curriculum System of Big Data Technology in Results-Oriented Vocational Education" (XHZDB2023006), Study on Strategies for Improving Teaching Quality for International Students (2023CXJG20), Jiangsu Provincial Education Science Planning "14th Five-year" project: Research on The Developmental Funding System of Higher Vocational Colleges based on Big Data (X-C/2020/06). Education Digitization: Construction of High-Quality Development System of Vocational Education(ZJ2023B197), Big Data Drives Innovation Research on Development-fund ed. Education in Higher Vocational Colleges (2022SJSZ0644), Research on Development-funded Education Innovation Driven by Big Data in Higher Vocational Colleges (2022SJSZ0644), Research Project of Institute of Changzhou University Higher Vocational Education: Teaching Reform Path of Computer Courses Empowered by Artificial Intelligence (CDGZ2023029).

References

1. Chunli, Z.: Significance, value and path of the construction of professional teaching resource base in vocational education[J]. Vocat. Educ. Forum. **36**(10), 58–62 (2019)
2. Xue-Zhi, L., Li-shuang, F.: Personalized learning in Japanese primary and secondary education: experience, problems and enlightenment[J]. Comp. Educ. Res. **02**, 13–17 (2006)
3. Wenhui, P.: Network Learning Behavior Analysis and Modeling[M]. Science Press, Beijing (2013)
4. Jinglan, J.: Construction and Application of Learning Style Data Mining Model Based on Network Learning Behavior Model[M]. Northwestern Polytechnical University Press, Shaanxi (2018)
5. Bandura.: Social Foundation of Thought and Action: Social Cognition[M]. East China Normal University Press, Shanghai (2018)

Chapter 9
Optimizing Deep Learning Networks for Image Classification Using the Rafflesia Optimization Algorithm

Shu-Chuan Chu, Ruhai You, Jeng-Shyang Pan, Lingping Kong, and Chin-Shiuh Shieh

Abstract Hyperparameters play a crucial role in neural networks, affecting the performance and generalization ability of the model. However, manual tuning of the parameters is a difficult and time-consuming task because the interactions and effects between the hyperparameters are not easily captured. ResNet is a classical neural network that has been widely used in image classification tasks, but its performance can be further improved by optimizing the hyperparameters. In this chapter, we propose a method to automatically adjust the hyperparameters by optimizing the two hyperparameters of the ResNet using the Dahlia algorithm to find the optimal configurations in the hyperparameter search space. Through the search and selection process of the Rafflesia optimization algorithm (ROA), we successfully find the hyperparameter configurations with the optimal classification accuracies on the CIFAR10 dataset. The experimental results show that compared with the traditional ResNet using the default hyperparameters, the method proposed in this chapter achieves a significant improvement on the CIFAR10 dataset and exceeds the performance of other common networks.

S.-C. Chu (✉) · R. You (✉)
College of Computer Science and Engineering, Shandong University of Science and Technology, Qingdao, China
e-mail: scchu0803@sdust.edu.cn; youruhai@sdust.edu.cn

J.-S. Pan
College of Computer Science and Engineering, Shandong University of Science and Technology, Qingdao, China

Department of Information Management, Chaoyang University of Technology, Taichung, Taiwan

L. Kong
VŠB – Technická univerzita Ostrava, Faculty of Computer Science, Ostrava, Czechia
e-mail: lingping.kong@vsb.cz

C.-S. Shieh
Department of Electronic Engineering, National Kaohsiung University of Science and Technology, Kaohsiung, Taiwan

S.-P. Tseng et al. (eds.), *Advances in Intelligent Information Hiding and Multimedia Signal Processing, Volume 2*, Smart Innovation, Systems and Technologies 416, https://doi.org/10.1007/978-981-97-8760-9_9

9.1 Introduction

Image classification is an important task in the field of computer vision [1–3], and deep learning algorithms [4–6] have made significant breakthroughs in this task. Convolutional neural network (CNN) is one of the most commonly used and effective deep learning algorithms for image classification task. LeNet [7, 8] proposed by Yann LeCun et al. in 1998 was the first convolutional neural network to achieve success in image classification task. LeNet was initially actualized for handwritten digit recognition task and contains convolutional, pooling, and fully connected layers; AlexNet [9], proposed in 2012 by Alex Krizhevsky et al., was a notable breakthrough in the ImageNet Image Classification Challenge. AlexNet reduces overfitting through the use of the rectified linear unit (ReLU) as an activation function and the dropout technique. Its success has brought deep learning to the forefront of computer vision. VGGNet [10], proposed by Simonyan and Zisserman in 2014, improves performance by using relatively small convolutional kernels (3x3) and a deeper network architecture. VGGNet employs multiple repeating blocks of convolutional layers and fully connected layers, which allows for more expressive and better feature extraction. VGGNet uses multiple repeating blocks of convolutional layers and fully connected layers and better feature extraction. GoogLeNet [11] was proposed by Google team in 2014 with the introduction of the inception module. The inception module employs multiple convolutional layers operating in parallel in order to extract features at different scales, thus effectively reducing the number of parameters. ResNet [12] is a network proposed by Kaiming He et al. in 2015, which solves the problem of deep networks by introducing residual connections to solve the problem of gradient vanishing and gradient explosion in deep networks. Residual connectivity allows information to skip directly from one layer to another, allowing the network to be deeper, and has been a huge success in the ImageNet Image Classification Challenge [13, 14].

CNN architectures have achieved great success in image classification tasks; these architectures are based on layer-based feature extraction with hyperparameter values that determine the effectiveness of the CNN model, finding the optimal values of these hyperparameters can be painstaking, time-consuming, and with various trial and errors. Hyperparameter tuning is considered to be a Non - Deterministic Polynomial -hard optimization problem with respect to a vast search space, and meta-heuristic algorithms are considered to make a very effective method for solving such problems [15, 16].

Metaheuristic algorithms are a class of optimization algorithms based on heuristic search strategies for the optimization or near optimization of problems [17, 18]. These algorithms draw on phenomena such as evolution, group behavior, and physics in nature to search the solution space and optimize it progressively through iterative and adaptive mechanisms. In order to solve optimization problems in different fields, researchers have proposed many new optimization algorithms such as cat swarm optimization (CSO) [19] algorithm, fish migration optimization (FMO) [20] algorithm, gray wolf optimization (GWO) [21] algorithm, whale optimization algorithm (WOA), butterfly optimization algorithm (BOA) [22], and recently proposed gannet optimization algorithm (GOA) [23]. After these algorithms were proposed, their results were applied to problems such as feature selection, image segmentation, wireless sensor node localization, and engineering optimization [24, 25].

The Rafflesia Optimization Algorithm (ROA) [26], implemented by Pan et al., is an efficient meta-heuristic algorithm, an intelligent evolutionary algorithm based on Rafflesia's life habit. It consists of three main phases: attracting insects, swallowing insects, and dispersing seeds. The ROA algorithm proves its effectiveness by comparing it with other well-known algorithms on 23 functions and IEEE CEC 2013 benchmarks [27]. Its versatility has been demonstrated by its use in solving a wide range of engineering problems [28].

9.2 Related Work

9.2.1 ResNet

Residual network is a very famous and important deep learning model [29, 30]. The core idea of ResNet is to introduce residual connections to solve the degeneration problem of neural networks, allowing the network to learn better as the number of layers increases. Traditional neural network models face the problem of gradient vanishing and gradient explosion when the number of layers increases, resulting in the network not being trained properly. To solve this problem, ResNet proposes the concept of residual block. In a residual block, the identity mapping of the input is added to the residual function through jump connections, allowing the network to learn the residuals directly instead of relearning the entire function. This design allows the network to easily fit the identity mapping and thus train the deep network more efficiently. As shown in Fig. 9.1. The complete ResNet18 network structure used in this chapter is shown in Fig. 9.2.

9.2.2 Rafflesia Optimization Algorithm (ROA)

Rafflesia is a saprophytic plant. It has no leaves or underground stems and an exceedingly short main axis. When it blooms, it emits a distinctive odor that attracts pollinators. Some of the attracted insects may become stuck and die during the pollination process as a result of its distinctive structure. The flower's petals wither after the blossoming phase, and the fruit becomes mature at that time. Many tiny seeds will be present in the fruit. When the fruits fall to the ground, the seeds are randomly distributed to numerous locations in various ways in order to find a

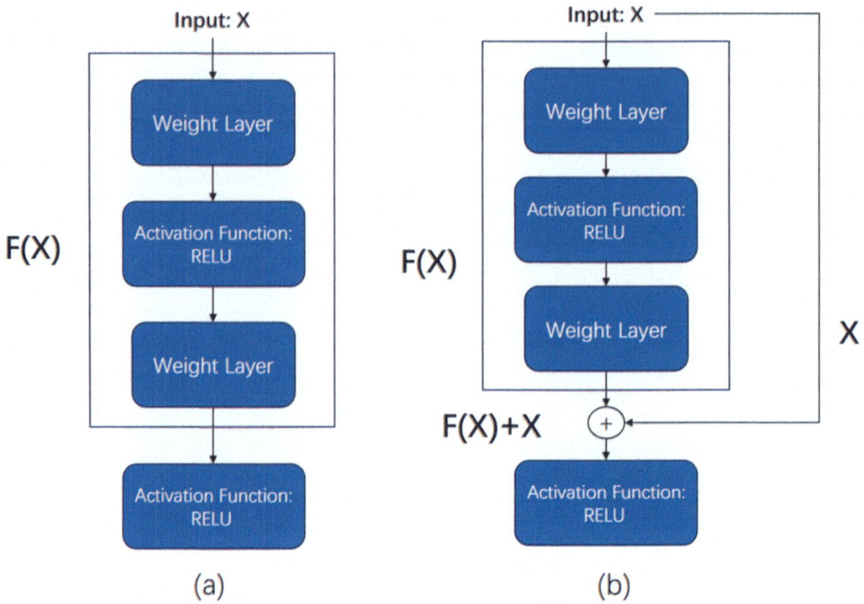

Fig. 9.1 ResNet group blocks: (**a**) the plain block and (**b**) the residual block

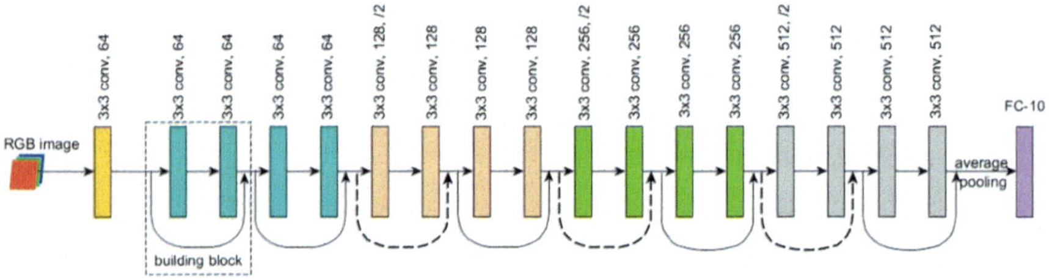

Fig. 9.2 Deep ResNet18 architecture

good location for germination. This study realizes ROA in three stages based on the characteristics listed above.

The first stage is the attraction of insects, which is divided into two strategies. The first strategy is to replace inferior individuals. In this strategy, one third of the inferior individuals are replaced by new individuals. Each dimension of the newly added individual is abstracted into a three-dimensional space for computation.

The equations used to compute the position of individual X are (9.1)–(9.3) to update its position.

$$X_{i_k} = X_{\mathrm{best_k}} + d \cdot \sin \beta_k \cos \gamma_k \tag{9.1}$$

$$d = \sqrt{\sum_{k=1}^{D} \left(X_{\gamma_k} - X_{\text{best}_k}\right)^2} \tag{9.2}$$

$$X_{\text{worst}_i} = X_i \tag{9.3}$$

where d is the distance between X_i and X_{best}, X_R is the best individual in the group, X_{worst} denotes the inferior individual in the group, β_k is $[0, \pi/2]$, and γ_k is a random value between $[0, \pi]$.

The second technique involves updating two thirds of the better-adapted individuals, with the individual velocity updating equations generated from the investment flatting flying model. Individual velocity-updating Eqs. (9.4)–(9.7) are illustrated, and the individual updating equation is (9.8):

$$\overrightarrow{v_1} = \frac{\omega_0}{2}\sqrt{A^2\sin^2(\omega_0 t + \theta) + B^2\cos^2(\omega_1 t + \theta)} \tag{9.4}$$

$$\overrightarrow{v_2} = \overrightarrow{v_2}\omega_0 \cos(\omega_0 t + \theta + \varphi) \tag{9.5}$$

$$\overrightarrow{v} = \overrightarrow{v_1} + \overrightarrow{v_2} \tag{9.6}$$

$$\overrightarrow{le} = C \cdot \overrightarrow{v} \cdot t + \left(X_{\text{best}} - X_{(t)}\right) \cdot (1 - C) \cdot \text{rand} \tag{9.7}$$

$$X_{(t)} = X_{(t)} + \overrightarrow{le} \tag{9.8}$$

where $\overrightarrow{v_1}$ and $\overrightarrow{v_2}$ denote the translational and rotational velocities, respectively. w_0 and w_1 denote the frequency period of flutter and lateral flutter, respectively, both of which have a value of 0.025. A is the amplitude of the wing during motion and has a value of 2.5. B is the lateral offset and has a value of 0.1. ϕ denotes the phase difference between translation and rotation and has a value of -0.78545. The value of θ is in the range of $(0, 2\pi)$. The initial value of V_2 is in the range $(0, 2\pi)$. C is a random number in $(-1,1)$.

The worm swallowing step eliminates the poorly adapted individuals after a set number of iterations, maintaining the quality of the comprehension and increasing the algorithm's efficiency.

During the seeding phase, the location of the dominants is represented by the initial optimal individual for that phase. Meanwhile, other individuals randomly look around for a suitable environment for growth. In this stage, the individual updating equation is as follows:

$$X_{(t)_k} = X_{\text{best}_k} + rd \cdot \exp\left(\frac{\text{item}}{\text{Max}_{\text{iter}}} - 1\right) \cdot \text{sign}(\text{rand} - 0.5) \tag{9.9}$$

$$rd = \text{rand}\cdot(ub - lb) + lb \tag{9.10}$$

where ub is the upper bound of the search space and lb is the lower bound of the search space.

9.3 Proposed Model: Hybrid ROA-ResNet18

This section introduces a new integration of ROA and ResNet18 for image classification, based on hyperparameter optimization. To achieve accuracy, the ROA algorithm selects the optimal hyperparameter values for ResNet18. After setting the best hyperparameters, the network is trained and tested on the CIFAR10 dataset. We employ ROA to fine-tune two hyperparameters of ResNet18, namely batch size and learning rate. Therefore, the search space is two dimensional, where each position in the search region corresponds to a set of these two hyperparameters.

The ROA algorithm optimizes the hyperparameters of ResNet18 through an adaptive searching strategy. In the initial search phase, ROA randomly selects a set of hyperparameters as the starting point and utilizes the residual mean strategy to update candidate solutions. Subsequently, a combination of local search and global search is employed by ROA to iteratively update candidate solutions and gradually improve performance. During the search process, the ROA algorithm evaluates the performance of different hyperparameter combinations using evaluation metrics such as cross-entropy loss and selects the solution with the best performance as the final result.

The search space for ResNet18 fine-tuning is specified as a two-dimensional plane with one dimension representing the batch size and the other dimension representing the learning rate. Each position in the search region corresponds to a set of batch size and learning rate values. ROA explores different positions in the search region and evaluates the performance of different combinations. By selecting the hyperparameter combination with the best performance, the optimal hyperparameter configuration for ResNet18 is determined.

Once the optimal hyperparameters are determined, ResNet18 is trained and tested on the CIFAR10 dataset. By utilizing the ROA-optimized hyperparameters, the network is better adapted to the requirements of the image classification task, leading to improved classification accuracy. In conclusion, through this new integration of ROA and ResNet18, we expect to achieve better performance and results in image classification tasks.

Table 9.1 Settings of the parameter values for ROA in combination with ResNet50

Heading level	Example
Max_iter	20
Pop_size	30
Dimension	2
Learning rate	[1e-7, 1e-3]
Batch size	[64, 512]
Training epochs of ResNet50	10

Table 9.2 The hyperparameters' optimal values as determined by ROA

Hyperparameter	Best value
Batch size	127
Learning rate	0.0008

9.4 Experimental Evaluation and Discussion

Table 9.1 reports the settings of the parameter values for the combination of the ROA algorithm and the ResNet18 model. The table includes the following hyperparameters and their search spaces:

Max_iter: this hyperparameter specifies the maximum number of iterations to run the optimization algorithm. Pop_size: this hyperparameter determines the population size, which is the number of candidate solutions in each iteration of the optimization algorithm. It is set to 30, meaning that 30 candidate solutions will be evaluated in each iteration. Dimension: this hyperparameter specifies the dimension of the search space, i.e., the number of hyperparameters to be optimized. Learning rate: this hyperparameter controls the step size of the optimization algorithm to update the neural network weights during the training period. The search space for this hyperparameter is set to [1e-7, 1e-3]. Batch size: this hyperparameter specifies the number of samples used in each iteration during training, and the search space for this hyperparameter is set to [64, 512]. Also, the training process of the ResNet18 model involves ten epochs during the search process.

Algorithm 9.1: ROA-ResNet18

The goal of the ROA algorithm is to improve the accuracy of the model on the test set as much as possible, and by evaluating the accuracy of the test set using these parameters, the desired values of batch size and learning rate have been chosen. Table 9.2 lists the optimal hyperparameters determined using the ROA algorithm. XX accuracy has been achieved on CIFAR10. A comparison of the accuracy of ROA-ResNet18 with other models is presented in Table 9.3.

Input: *f(x)*: fitness function; *N*: individual number; *d*: function dimension; *lb*: maximum boundary; *ub*: minimum boundary; *Max_iter*: maximum number of iteration
Output: global optimal value.
Initialize the related parameters of ROA Randomly
generate the positions of the insects
That is, the value of the initialization hyperparameter
while iter<*Max_iter* **do**
  Calculate the fitness of insects. Setting the hyperparameters to train and validate the model Sort individuals according to fitness, and update the optimal individual X. **if** *rem(iter,pd)<(pd/2)* **then**
    for *i=1:N* **do**
      if *i<=(N/3)* **then**
        Update the individuals according to equations (1)-(3)
      else
        Update the individuals according to equations (4)-(8)
  if *rem(iter,pd)==(pd/2)* **then**
    Stage of swallowing insects,delete the worst fitness individual.
  if *rem(iter,pd)>(pd/2)* **thenfor**
    i=1:N **do**
      Use (10) to calculate the value of rd.
      Update the individual with (9).

Table 9.3 Model accuracy comparison

Model	Accuracy
LeNet	80%%
VGGNet	88%
GoogleNet	85%
Vit	80%
MLP	88%
ResNet18	91%
ROA-ResNet18	94%

9.5 Conclusions and Future Work

In this chapter, a reliable method is proposed to automatically search the hyperparameters of neural networks, and the hyperparameters of ResNet are optimized by using the Rafflesia optimization algorithm (ROA), which can significantly improve the performance of the network. Compared with the traditional manual parameter tuning method, the ROA algorithm can comprehensively explore the hyperparameter space through the self-adaptive searching strategy, which makes the network better adapt to the specific characters' requirements. The results show that the ResNet optimized by the Rafflesia optimization algorithm achieves better performance in the image classification task.

Second, ResNet optimized by hyperparameters exhibits higher accuracy and lower training loss in image classification tasks. Compared with the unoptimized default parameters, the model optimized by the Dahlia algorithm can fit the training dataset better and improve the classification accuracy under the same training conditions. This demonstrates the effectiveness and feasibility of the Rafflesia algorithm in optimizing the ResNet hyperparameters.

The experiments in this chapter are only for the image classification task, and in the future, the Daiwanghua algorithm can be applied to other computer vision tasks, such as target detection and semantic segmentation to further verify its generality and superiority. Second, the search strategy and hyperparameter settings in the Dahlia algorithm can also be improved. The search efficiency and classification performance of the algorithm can be further improved by introducing more search strategies, changing the search space or adding heuristic operators.

In summary, the hyperparameter optimization of ResNet by the Dahlia algorithm and its application to the image classification task have achieved satisfactory results in the experimental results. Future research work can further extend the application scope of the algorithm, improve the search strategy, and optimize the efficiency of the algorithm to achieve wider and more efficient applications.

References

1. Nguyen, T.-T., Nguyen, T.D., Nguyen, V.-T.: An optimizing pulse coupled neural network based on golden eagle optimizer for automatic image segmentation. J. Inf. Hiding Multimed. Signal. Process. **13**, 155–164 (2022)
2. Dengsheng, L., Weng, Q.: A survey of image classification methods and techniques for improving classification performance. Int. J. Remote Sens. **28**(5), 823–870 (2007)
3. Zhou, L., Zhang, C., Liu, F., Qiu, Z., He, Y.: Application of deep learning in food: a review. Compr. Rev. Food Sci. Food Saf. **18**(6), 1793–1811 (2019)
4. Liao, L., Lin, J., Zhu, Y., Bi, S., Lin, Y.: A bi-direction LSTM attention fusion model for the missing poi identification. J. Netw. Intell. **7**(1), 161–174 (2022)
5. Chen, Y., Lin, Z., Zhao, X., Wang, G., Yanfeng, G.: Deep learning-based classification of hyperspectral data. IEEE J. Sel. Top. Appl. Earth Obs. Remote Sens. **7**(6), 2094–2107 (2014)
6. LeCun, Y., Bengio, Y., Hinton, G.: Deep learning. Nature. **521**(7553), 436–444 (2015)
7. LeCun, Y., Bottou, L., Bengio, Y., Haffner, P.: Gradient-based learning applied to document recognition. Proc. IEEE. **86**(11), 2278–2324 (1998)
8. Xiaoyu, X., Cui, J., Chen, X., Chen, C.-L.: A facial expression recognition method based on residual separable convolutional neural network. J. Netw. Intell. **7**(1), 59–69 (2022)
9. Krizhevsky, A., Sutskever, I., Hinton, G.E.: Imagenet classification with deep convolutional neural networks. Adv. Neural Inf. Proces. Syst. **25** (2012)
10. Simonyan, K., Zisserman, A.: Very deep convolutional networks for large-scale image recognition. arXiv preprint arXiv, *1409.1556* (2014)
11. Szegedy, C., Liu, W., Jia, Y., Sermanet, P., Reed, S., Anguelov, D., Erhan, D., Vanhoucke, V., Rabinovich, A.: Going deeper with convolutions. In: Proceedings of the IEEE conference on computer vision and pattern recognition, pp. 1–9 (2015)
12. He, K., Zhang, X., Ren, S., Sun, J.: Deep residual learning for image recognition. Proc. IEEE Conf. Comput. Vis. Pattern Recognit., 770–778 (2016)
13. Nath, S.S., Mishra, G., Kar, J., Chakraborty, S., Dey, N.: A survey of image classification methods and techniques. In: 2014 International Conference on Control, Instrumentation, Communication and Computational Technologies (ICCICCT), pp. 554–557. IEEE (2014)

14. Wang, W., Yang, Y., Wang, X., Wang, W., Li, J.: Development of convolutional neural network and its application in image classification: a survey. Opt. Eng. **58**(4), 040901–040901 (2019)
15. Emam, M.M., Samee, N.A., Jamjoom, M.M., Houssein, E.H.: Optimized deep learning architecture for brain tumor classification using improved hunger games search algorithm. Comput. Biol. Med. **160**, 106966 (2023)
16. He, K., Zhang, X., Ren, S., Sun, J.: Identity mappings in deep residual networks. In: Computer Vision–ECCV 2016: 14th European Conference, Amsterdam, The Netherlands, October 11–14, 2016, Proceedings, Part IV 14, pp. 630–645. Springer (2016)
17. Pan, J.-S., Liang, Q., Chu, S.-C., Tseng, K.-K., Watada, J.: A multi-strategy surrogate-assisted competitive swarm optimizer for ex- pensive optimization problems. Appl. Soft Comput. **147**, 110733 (2023)
18. Wong, W.K., Ming, C.I.: A review on metaheuristic algorithms: recent trends, benchmarking and applications. In: 2019 7th International Conference on Smart Computing & Communications (ICSCC), pp. 1–5. IEEE (2019)
19. Chu, S.-C., Tsai, P.-W., Pan, J.-S.: Cat swarm optimization. In: PRICAI 2006: Trends in Artificial Intelligence: 9th Pacific Rim International Conference on Artificial Intelligence Guilin, China, August 7–11, 2006 Proceedings 9, pp. 854–858. Springer (2006)
20. Pan, J.-S., Tsai, P.-W., Liao, Y.-B.: Fish migration optimization based on the fishy biology. In: 2010 Fourth International Conference on Genetic and Evolutionary Computing, pp. 783–786. IEEE (2010)
21. Mirjalili, S., Mirjalili, S.M., Lewis, A.: Grey wolf optimizer. Adv. Eng. Softw. **69**, 46–61 (2014)
22. Arora, S., Singh, S.: Butterfly optimization algorithm: a novel approach for global optimization. Soft. Comput. **23**, 715–734 (2019)
23. Pan, J.-S., Zhang, L.-G., Wang, R.-B., Snášel, V., Chu, S.-C.: Gannet optimization algorithm: a new metaheuristic algorithm for solving engineering optimization problems. Math. Comput. Simul. **202**, 343–373 (2022)
24. Pan, J.-S., Sun, B., Chu, S.-C., Zhu, M., Shieh, C.-S.: A parallel compact gannet optimization algorithm for solving engineering optimization problems. Mathematics. **11**(2), 439 (2023)
25. Houssein, E.H., Saad, M.R., Hashim, F.A., Shaban, H., Hassaballah, M.: Lévy flight distribution: a new metaheuristic algorithm for solving engineering optimization problems. Eng. Appl. Artif. Intell. **94**, 103731 (2020)
26. Pan, J.-S., Zonglin, F., Chia-Cheng, H., Tsai, P.-W., Chu, S.-C.: Rafflesia optimization algorithm applied in the logistics distribution centers location problem. J. Internet Technol. **23**(7), 1541–1555 (2022)
27. Liang, J.J., Qu, B.Y., Suganthan, P.N., Hernández-Díaz, A.G.: Problem definitions and evaluation criteria for the CEC 2013 special session on real-parameter optimization. Comput. Intell. Laborat. Zhengzhou University, Zhengzhou, China and Nanyang Technol. Univ. Singapore, Tech. Rep. **201212**(34), 281–295 (2013)
28. Yang, X.-S.: Optimization and metaheuristic algorithms in engineering. In: Meta- heuristics in Water, Geotechnical and Transport Engineering, vol. 1, p. 23 (2013)
29. Parvat, A., Chavan, J., Kadam, S., Dev, S., Pathak, V.: A survey of deep-learning frameworks. In: 2017 International Conference on Inventive Systems and Control (ICISC), pp. 1–7. IEEE (2017)
30. Tai, Y., Yang, J., Liu, X.: Image super-resolution via deep recursive residual network. Proc. IEEE Conf. Comput. Vis. Pattern Recognit., 3147–3155 (2017)

Chapter 10
Optimization of Wireless Sensor Networks Using Phasmatodea Population Evolution Algorithm

Jeng-Shyang Pan, Bing Sun, Shu-Chuan Chu, Xiaomin Liu, and Junzo Watada

Abstract In order to extend the lifetime of wireless sensor networks (WSNs) and their scalability, various methods have been proposed to solve this problem; one of the more effective methods is clustering. Clustering is an effective way to deal with topological problems, and it can be equally applied to the optimization of wireless wire sensor networks with network topology, where the choice of number and location of cluster heads in WSNs also has a significant impact on their lifetime as well as on various aspects of performance. In this chapter, the Phasmatodea population evolution algorithm is used for the selection of the number and location of cluster head nodes in WSNs to select the number and location of cluster heads that maximize the energy efficiency of the WSN network. It is shown experimentally that the proposed Phasmatodea population evolution algorithm–based optimization of WSN achieves good results.

J.-S. Pan
Qingdao, China

Taichung, Taiwan
e-mail: jengshyangpan@gmail.com

B. Sun · S.-C. Chu (✉)
College of Computer Science and Engineering, Shandong University of Science and Technology, Qingdao, China
e-mail: bingsun@sdust.edu.cn; scchu0803@sdust.edu.cn

X. Liu
The Heilongjiang Provincial Key Laboratory of Autonomous Intelligence and Information Processing, Jiamusi University, Jiamusi, China
e-mail: xiaominliu@vip.com

J. Watada
Graduate School of Information, Production and Systems, Waseda University, Kitakyushu, Japan

S.-P. Tseng et al. (eds.), *Advances in Intelligent Information Hiding and Multimedia Signal Processing, Volume 2*, Smart Innovation, Systems and Technologies 416, https://doi.org/10.1007/978-981-97-8760-9_10

10.1 Introduction

With the rapid development of industrial and scientific fields, which also demand the precision of technology, the wireless sensor network (WSN), a distributed network structure, has its energy consumption in all aspects of the communication process closely related to the selection of its network cluster—head nodes. WSN is widely used in various fields, such as detection of environmental parameters, military battlefield detection, agricultural soil moisture detection and real-time monitoring of logistics [1]. However, there are many problems in the application of WSN, such as energy constraints. Wireless sensor nodes are mainly powered by their own batteries, and the depletion of power energy leads to nodes that cannot work, thus affecting the performance and life of the network. In addition, there is a problem of the number of cluster head nodes, network size and so on. These problems can be solved by using some optimization algorithms to maximize the performance of WSNs in all aspects, and a proper selection of cluster head set can effectively manage the energy consumption and prolong the lifetime of the network. Various metaheuristic algorithms have been previously optimized for performance improvement in various aspects of WSNs, such as the particle swarm optimization algorithm (PSO) [2–4], flower pollination algorithm (FPA) [5, 6], bamboo forest growth optimization (BFGO) [7] and grasshopper optimization algorithm (GOA) [8–10]. Some improvement strategies of the algorithms have also emerged to enhance the performance of the algorithms [11–14].

In this chapter, we use the Phasmatodea population evolution (PPE) algorithm [15], which has been recently proposed in recent years to select the cluster points of WSNs and solve the energy optimization problem in WSNs. The bamboo worm optimization algorithm (PPE) is an algorithm inspired by the evolutionary behavior of bamboo worm populations in nature, which prolongs the survival cycle of WSNs through the bamboo worm optimization algorithm's management of energy and selection of the set of cluster points in WSNs. The PPE is different from other common algorithms in that the algorithm can virtually populate a single solution, which has two attributes, namely the number of the populations and the growth rate. The algorithm also incorporates elements of competition between populations and similar evolution. During the exploration of the virtual population of this algorithm, the size of its population is affected by it. In this chapter, PPE is compared with other algorithms with respect to the optimization problem of WSNs, and the results show that PPE improves the performance of WSNs more than other algorithms.

10.2 Basic Algorithm and Related Concepts

The PPE algorithm chosen in this chapter is based on the survival of bamboo worm populations in nature, where there are various interactions between populations and competitive behaviors between different populations [16, 17]. It is found that bamboo worm will evolve or compete with the change of population by the effect of size of bamboo worm population on its population in nature. The algorithm is proposed on the basis of competition between populations and growth of population size. When the algorithm is run, the particle update of the algorithm includes not only the initialization and the update of the position of the population but also the update of the evolutionary trend of the population during the evolutionary process [18, 19].

Suppose there are N_p solutions x in D-dimensional space, and x is used to represent a virtual population of bamboo worms, where $x = [x_1, x_2, \ldots, x_d]$.

In the initialization phase of the PPE algorithm, random generation is used to construct N_p populations for each solution x, where the formula for the number of populations for each solution x is given by

$$p_i = \frac{1}{N_p} \tag{10.1}$$

The population growth rate a_i, another attribute of each solution x, is initialized to 1.1. The evolutionary tendency of the population during the evolutionary process is defined as ev and initialized to $ev = 0$. After initializing each solution, the fitness can be computed for each x and the solution that is the best fitness is selected as the global optimal solution *Gbest* after initialization.

Based on the idea of population evolution, the bamboo worm algorithm will take a form resembling population evolution in nature through iterations during its execution. This can establish a connection between population evolution and the natural environment. It is important to note that the local optimum of the population does not necessarily represent the global optimum. According to this idea, the PPE algorithm chooses to utilize historical optimal solutions as references for guiding the next population evolution. In this algorithm, k historical optimal solutions are selected as references for adjusting the solutions within the population. These historical optimal solutions are stored in the set $Ho = [x_{h1}, x_{h2}, \ldots, x_{hi}, \ldots, x_{hk}]$, where i denotes the ith occurrence of an optimal solution, and x_{h2} is the optimal solution that occurs after the occurrence of x_{h1}. If the total number of x_h generated is N_p, then the number representation of k is

$$k = \log\left(N_p\right) + 1 \tag{10.2}$$

H_o is initialized by selecting k optimal solutions out of N_p solutions, and the corresponding fitness functions are stored in an array. With the completion of the population initialization, the PPE algorithm starts iterating. After each iteration, the position of the population is updated according to the evolutionary trend. The formula is

$$X^{t+1} = X^t + ev \tag{10.3}$$

In the above equation, t denotes the current number of iterations. After the completion of one round of iterations, the global optimal solution *Gbest* and k-optimal solution H_o need to be updated.

Population evolutionary trends are influenced by several aspects, including the evolutionary direction of the population itself, competition between populations and natural selection. There are two scenarios in terms of population evolution, with different evolutionary maneuvers based on the fitness of the updated population.

The first scenario is that the population becomes better adapted after one iteration in terms of fitness value than before; then this will have an impact on the current population size and growth rate. The change in the growth rate is

$$a^{it+1} = a^{it}\left(1 + \frac{f\left(x^t\right) - f\left(x^{t+1}\right)}{f\left(x^{t+1}\right)}\right) \tag{10.4}$$

In Eq. (10.4), if $f\ (x^{t+1})$ is better than $f\ (x^t)$, then an increase is made to the growth rate, at which point the evolutionary direction is favorable for the survival of the population. The effect on the population can be

$$P^{t+1} = a^{t+1} \cdot P^t\left(1 - p^t\right) \tag{10.5}$$

After the population size and growth rate have been updated, the evolutionary trend *ev* needs to be updated. The evolutionary trend *ev* is affected by three aspects, which are similar evolution, continuation of evolutionary trend and mutation within the population. Similar evolution is the updating of the position of the current solution toward the optimal solution closest to it. The continuation of the evolutionary trend is that if the fitness value of the solution after the move is better, then the trend of this move will be maintained. Mutations within the population increase the uncertainty of population evolution, and these mutations over many iterations can also affect the evolutionary trend. Based on these three aspects, the updated formula for the evolutionary trend *ev* is

$$ev^{t+1} = \alpha\left(1 - p^{t+1}\right)\left(s\left(Ho, x^t\right) - x^t\right) + p^{t+1}ev^t + \beta p^{t+1}m \tag{10.6}$$

The second case is the opposite of the first; if the fitness value becomes worse after the population is moved, there is some probability of accepting this worse solution. A uniform distribution is used to generate values from 0 to 1, and then the acceptance probability p is set, and if it is within the acceptance probability p, then the worse position is accepted; otherwise it is not accepted. The updated evolutionary trend formula is

$$ev^{t+1} = \left(s\left(Ho, x^t\right) - x^t\right) \cdot \text{rand} + st \cdot rdn \tag{10.7}$$

where *st* is used to control the extent of the area around the exploration, calculated by the formula

$$st^{t+1} = \left(\frac{1}{t}\right)^{\left|\frac{f(x^{t+1})-f_{\min}^{t+1}}{f_{\min}^{t+1}-f_{\max}^{t+1}}\right|} \cdot c^{t+1} \tag{10.8}$$

where c is initialized to a constant 2 and slowly gets smaller with iterations; after that, the update formula is $c^{t+1} = c^t \times 0.99$.

The competition that exists between populations also has an effect on the evolutionary trend of the population, and after the above updating steps, the updated equation for the effect of population competition on the population is

$$p_i = p_i + a_i p_i \left(1 - p_i - \frac{f(x_j)}{f(x_i)} p_j\right) \tag{10.9}$$

When competition occurs between the populations of x_i and x_j, the population size of x_i is affected by the competition, and the updating formula of its size is shown above. After the population size changes, the evolutionary trend of the population should also be updated, and the updating formula is

$$ev^{t+1} = ev^{t+1} + \frac{f(x_j) - f(x_i)}{f(x_j)} (x_j - x_i) \tag{10.10}$$

In this chapter, we use G, which denotes the distance between two populations, to determine whether or not competition has occurred between two populations. If the distance between two populations is less than G, it means that there is going to be competition between the two populations, where G is calculated as $G = 0.1 \cdot (ub - lb)((Maxg + 1 - it)/\text{Max}g)$. Max$g$ is the maximum number of iterations. Equations (10.9) and (10.10) were then used to calculate the competition results.

To better optimize the WSN network, this chapter adds a perturbation strategy to the PPE algorithm. In order to allow the population to better jump out of the local optimal solution, the iteration number perturbs the particles of the current iteration every ten generations [20].

Combining the above descriptions, the PPE-WSN pseudo-code is shown in Algorithm 10.1.

10.3 Energy Modeling for WSNs

In infinite sensor network, in order to collect the data from the surrounding environment, several number of nodes needs to be distributed to aggregate and organize the data from the surrounding. Cluster head nodes are the bridge to establish data transmission between each non-cluster head node and BS, so the cluster head nodes consume more energy. The life cycle of WSN is affected by various conditions, such as the selection and deployment of the location of cluster head nodes, the location of the BS, and the environmental area in which the individual nodes are located; so the selection of the appropriate cluster head in a WSN is an important factor to extend its life cycle, and using optimization algorithms to solving this problem is currently the more effective method [21]. The basic schematic diagram of WSN model is as follows:

In the WSN model of this chapter, the energy loss during transmission at each node is formulated as follows:

$$\begin{cases} E_{CH}^{T} = Lp \times (Et + Ea) + Lp \times E_{am} \times dis^2 \\ E_{Non}^{T} = Lp \times Et + Lp \times E_{am} \times dis^2 \end{cases} \tag{10.11}$$

where E_t denotes the energy lost in transmitting a data unit, L_p is the length of the packet, *dis* denotes the distance over which the message is transmitted, *Non* E_{am} denotes the parameter of the amplifier circuit model, and E^T and E_{CH} denote the energy loss of non-cluster head and cluster head nodes, respectively, during the transmission of the message. The following equation represents the energy lost by the cluster head node in receiving and aggregating the messages from non-cluster head nodes (Fig. 10.1).

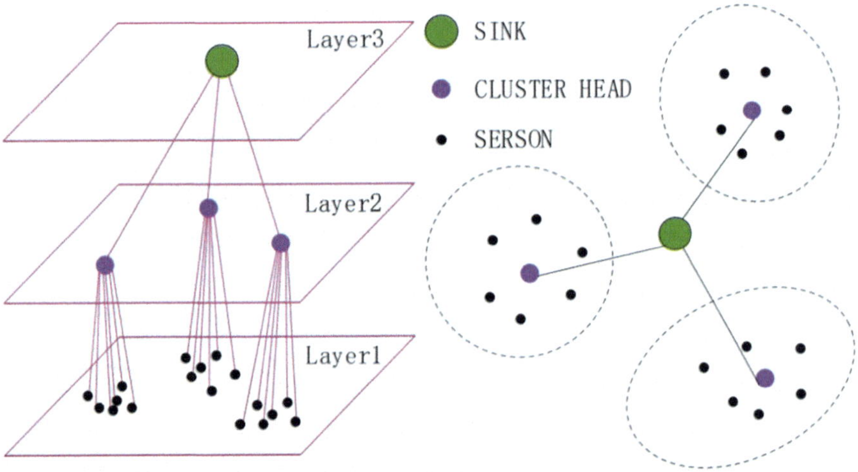

Fig. 10.1 Schematic diagram of the relationship between the nodes of the WSN

Algorithm 10.1: PPE-WSN

```
Input: clusterModel, clusterFunParam
Output: clusterModel
Initialize the relevant parameters of the PPE, including Np,ev,p,k and a;
Randomly generate network nodes in the given space and initialize the BS
position of the WSN;
Calculate the fitness value f(x) of the initialized WSN network, set Gbest and
 Ho;
for t = 2 to Maxgen do
    Update each x to newx using equation (3);
    Calculate new fitness newx, update Gbest and Ho;
    if t mod 10 = 0 then
        Perform perturbation updating of populations
    for i = 1 to Np do
        Update ai and pi using equations (4) and (5);
        if f(newx) ≤ f(x) then
            Update x, x = newx, update f(x);
            Update evi using equation (6);
        if f(newx) > f(x) then
            if rd < pi then
                Update x, x = newx, update f(x);
                Update evi using equations (7) and (8);
                Randomly choose a solution xj (j ≠ i);
                if dist(xj, xi) < G then
                    Update pi using equation (9), update evi using equation
                     (10);
                if pi ≤ 0 or ai ≤ 0 or ai > 4 then
                    Eliminate x and reinitialize;
```

$$E^R = L_p \times (E_r + E_a) \tag{10.12}$$

where E_r denotes the energy consumed to receive a unit byte message and E_a is the energy consumed to aggregate a unit byte message. The fitness function used in WSN to evaluate the set of selected cluster head nodes is defined in Eq. (10.13). The total energy consumption for this round of routing is

$$\text{Fitness} = E_{\text{total}} = \sum_{i=1}^{K_n} E_{\text{Non}}^T + \sum_{i=1}^{M_n} \left(E_{CH}^T + E^R \right) \tag{10.13}$$

where M denotes the number of cluster head nodes to be selected, and the number of non-cluster head nodes is K. The first and second parts of the equation denote the energy lost from the informative non-cluster head node to the cluster head node and the cluster head node to the base station, respectively.

10.4 Simulations and Results

In the whole wireless sensor network, the location and number of sensor nodes, the distribution of their location in the target area, the initial energy, and the location of the BS of the base station have a great impact on the energy consumption and the survival period in the WSN. In order to test the effect of PPE on the selection of the most available cluster head set for WSNs, this chapter builds a wireless sensor network model and compares it with the classical routing algorithm low-energy adaptive clustering hierarchy (LEACH) [22], the intelligent optimization algorithm particle swarm optimization-clustering (PSO-C) [23] and hybrid firefly particle swarm optimization (HFPSO) [24]. To ensure the fairness of the comparison, the number of iterations of each algorithm is set to 20.

Figure 10.2 shows the basic layout of the wireless sensor network model including the location of the BS, the location of each sensor node and the area of the network, where the area of the network for WSN is set as 200*200 m. The relevant parameters of the WSN model are shown in Table 10.1, where P_{CH} denotes the probability of a sensor node being selected as a cluster head.

In Fig. 10.3, this chapter shows the comparison between PPE algorithm and other three algorithms for WSN optimization with respect to the survival cycle. As the number of dead nodes increases, the survival cycle of WSN is coming to an end. When the number of dead nodes reaches the number of initialized is nodes, it means the survival cycle of WSN is over. From the results, it can be seen that PPE is the best

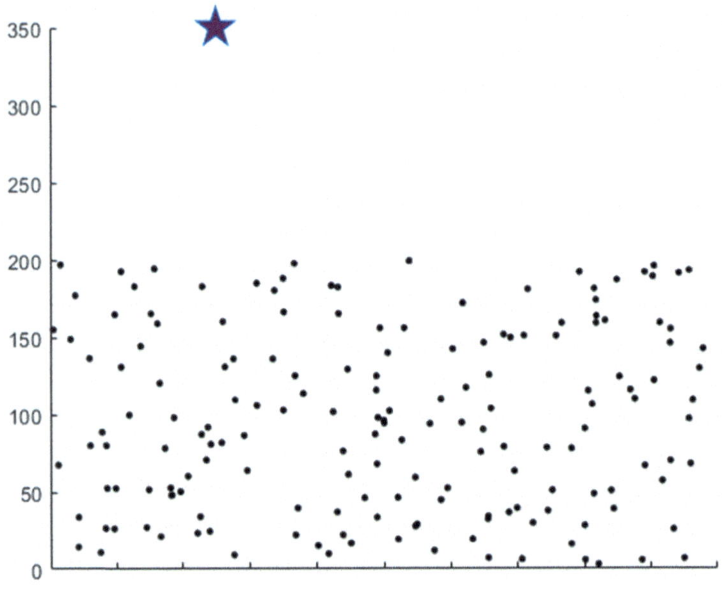

Fig. 10.2 Deployment diagram of the WSN model

Table 10.1 The parameters of the WSN model

Node initial energy	Base station location	Number of sensors	P_{CH}
0.5 J	(50, 350)	100	8%
L_p	6000 bit		
E_t	5×10^{-8} J		
E_r	2×10^{-8} J		
E_a	5×10^{-9} J		
E_{am}	1×10^{-10} J		

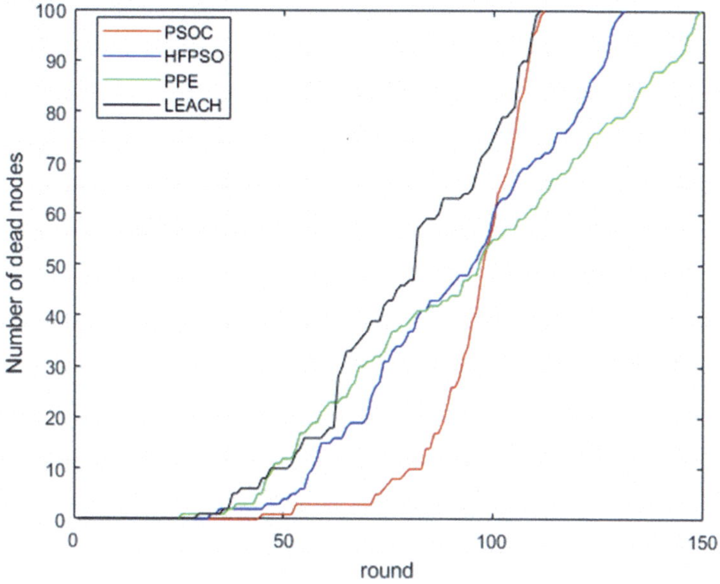

Fig. 10.3 Number of node deaths

for optimizing the WSN and making its life cycle longer by 33%, 13.7% and 34.2% compared to PSO algorithm, HFPSO algorithm and LEACH algorithm, respectively. From the energy consumption Fig. 10.4, it can be seen that the energy consumption of the sensor network of the PPE algorithm is not significantly different from the other algorithms in the early stage. However, the average energy consumption of the sensors optimized by the PPE algorithm in the later stages has been greater than the average energy consumption of the sensors optimized by the other algorithms. This is why the WSN optimized by the PPE algorithm can have a longer life cycle than the WSN optimized by other algorithms.

Table 10.2 shows the comparison of PPE algorithm with other algorithms in optimizing the WSN network in various aspects. Figs. 10.5 and 10.6 show the change in the amount of information received by the PPE algorithm versus other algorithms in cluster head (CH) and base station (BS) as the rounds progress. It can

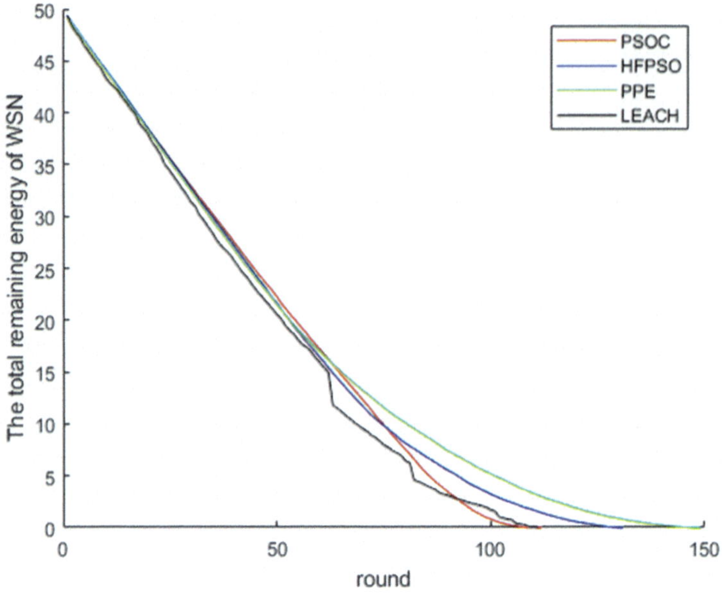

Fig. 10.4 Energy consumption of WSNs

Table 10.2 Comparison of data after optimization of WSN by algorithms

Algorithm	Life cycle (round)	Messages received	
		BS	CH
PPE	149	835,478	76,166
PSO	112	562,020	49,603
HFPSO	131	693,024	62,287
LEACH	111	498,417	43,576

be found that the PPE-optimized WSN used in this chapter is better in extending the life cycle of the WSN and receiving more data information.

10.5 Conclusions and Future Work

The results presented in this chapter for optimizing WSNs using the PPE algorithm compared with the classical routing algorithm LEACH algorithm, the swarm intelligence optimization algorithm PSO and HFPSO show that the optimization of WSNs using the PPE algorithm is superior to the other algorithms.

Simulation results show that utilizing the PPE algorithm effectively improves the energy efficiency and survival period of the WSN while maintaining a feasible packet throughput. In the future research direction of WSN, an intelligent optimization algorithm with multiple strategies can be developed to optimize more aspects

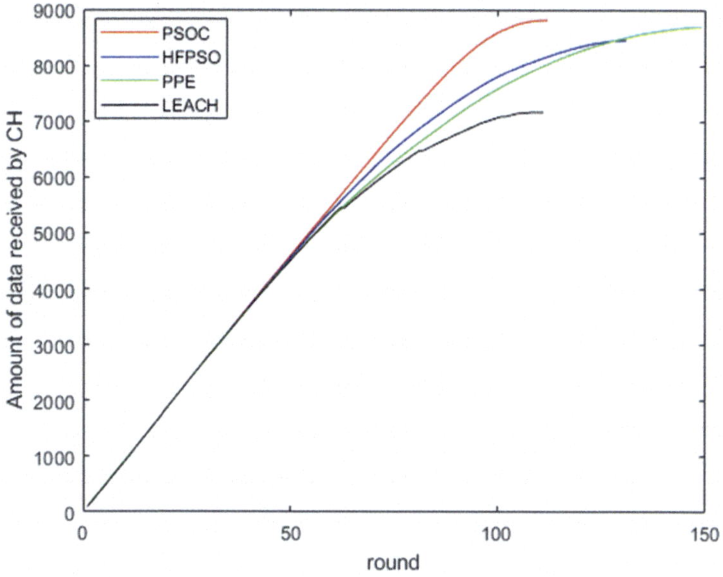

Fig. 10.5 The amount of messages received by CH

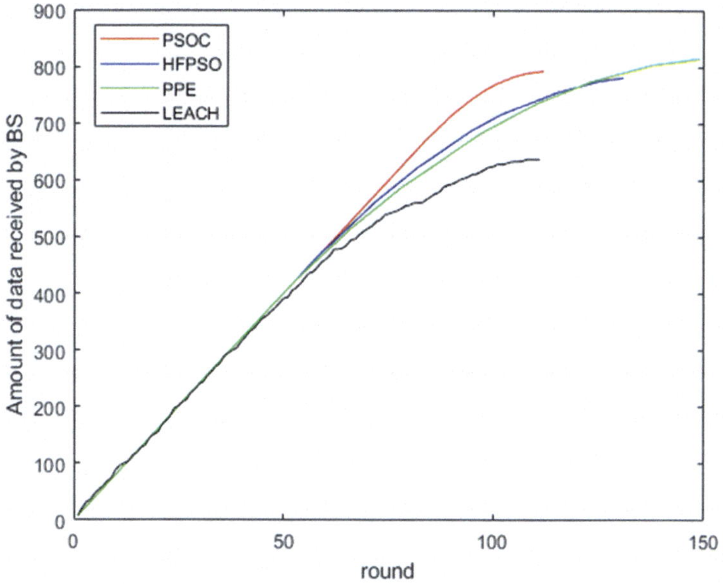

Fig. 10.6 The amount of information received by BS

of the wireless sensor network, which can better utilize the energy of the sensors and improve the energy efficiency of the WSN while maintaining a certain survival period [25–27]. The proposed method may be further improved by adopting some advanced algorithms [28–30].

References

1. Akyildiz, I.F., Weilian, S., Sankarasubramaniam, Y., Cayirci, E.: A survey on sensor networks. IEEE Commun. Mag. **40**(8), 102–114 (2002)
2. Kennedy, J., Eberhart, R.: Particle swarm optimization. In: Proceedings of ICNN'95-International Conference on Neural Networks, vol. 4, pp. 1942–1948. IEEE (1995)
3. Marini, F., Walczak, B.: Particle swarm optimization (pso). A tutorial. Chemom. Intell. Lab. Syst. **149**, 153–165 (2015)
4. Jain, M., Saihjpal, V., Singh, N., Singh, S.B.: An overview of variants and advancements of pso algorithm. Appl. Sci. **12**(17), 8392 (2022)
5. Yang, X.-S.: Flower pollination algorithm for global optimization. In: International Conference on Unconventional Computing and Natural Computation, pp. 240–249. Springer (2012)
6. Abdel-Basset, M., Shawky, L.A.: Flower pollination algorithm: a comprehensive review. Artif. Intell. Rev. **52**, 2533–2557 (2019)
7. Chu, S.-C., Feng, Q., Zhao, J., Pan, J.-S.: Bfgo: bamboo forest growth optimization algorithm. J. Internet Technol. **24**(1), 1–10 (2023)
8. Meraihi, Y., Gabis, A.B., Mirjalili, S., Ramdane-Cherif, A.: Grasshopper optimization algorithm: theory, variants, and applications. IEEE Access. **9**, 50001–50024 (2021)
9. Mirjalili, S.Z., Mirjalili, S., Saremi, S., Faris, H., Aljarah, I.: Grasshopper optimization algorithm for multi-objective optimization problems. Appl. Intell. **48**, 805–820 (2018)
10. Abualigah, L., Diabat, A.: A comprehensive survey of the grasshopper optimization algorithm: results, variants, and applications. Neural Comput. & Applic. **32**(19), 15533–15556 (2020)
11. Xi, J., Chen, Y., Liu, X., Chen, X.: Whale optimization algorithm based on nonlinear adjustment and random walk strategy. J. Netw. Intell. **7**(2), 306–318 (2022)
12. Neri, F., Mininno, E., Iacca, G.: Compact particle swarm optimization. Inf. Sci. **239**, 96–121 (2013)
13. Pan, J.-S., Zhu, M., Chu, S.-C., Snášel, V., Kong, L.: Robust digital watermarking with parallel compact sparrow search algorithm applied for QR code. J. Inf. Hiding Multim. Signal Process. **13**(2), 124–144 (2022)
14. Abdel-Basset, M., Mohamed, R., Abouhawwash, M.: A new fusion of whale optimizer algorithm with Kapur's entropy for multi-threshold image segmentation: analysis and validations. Artif. Intell. Rev. **55**(8), 6389–6459 (2022)
15. Song, P.-C., Chu, S.-C., Pan, J.-S., Yang, H.: Phasmatodea population evolution algorithm and its application in length-changeable incremental extreme learning machine. In: 2020 2nd International Conference on Industrial Artificial Intelligence (IAI), pp. 1–5. IEEE (2020)
16. Bedford, G.O.: Biology and ecology of the phasmatodea. Annu. Rev. Entomol. **23**(1), 125–149 (1978)
17. Yang, H.R., Shi, C.F., Engel, M.S., Zhao, Z.P., Ren, D., Gao, T.P.: Early specializations for mimicry and defense in a jurassic stick insect. Natl. Sci. Rev. **8**, nwaa056 (2021)
18. Karaboğa, D., Ökdem, S.: A simple and global optimization algorithm for engineering problems: differential evolution algorithm. Turk. J. Electr. Eng. Comput. Sci. **12**(1), 53–60 (2004)
19. Kai Qin, A., Huang, V.L., Suganthan, P.N.: Differential evolution algorithm with strategy adaptation for global numerical optimization. IEEE Trans. Evol. Comput. **13**(2), 398–417 (2008)
20. Pan, J.-S., Sun, B., Chu, S.-C., Zhu, M., Shieh, C.-S.: A parallel compact gannet optimization algorithm for solving engineering optimization problems. Mathematics. **11**(2), 439 (2023)

21. Jian, W., Ming, X., Liu, F.-F., Huang, M., Ma, L., Zhe-Ming, L.: Solar wireless sensor network routing algorithm based on multi-objective particle swarm optimization. J. Inf. Hiding Multim. Signal Process. **12**(1), 1–11 (2021)
22. Heinzelman, W.R., Chandrakasan, A., Balakrishnan, H.: Energy-efficient communication protocol for wireless microsensor networks. In: Proceedings of the 33rd Annual Hawaii International Conference on System Sciences, p. 10. IEEE (2000)
23. Abdul Latiff, N.M., Tsimenidis, C.C., Sharif, B.S.: Energy-aware clustering for wireless sensor networks using particle swarm optimization. In: 2007 IEEE 18th International Symposium on Personal, Indoor and Mobile Radio Communications, pp. 1–5. IEEE (2007)
24. Pitchaimanickam, B., Murugaboopathi, G.: A hybrid firefly algorithm with particle swarm optimization for energy efficient optimal cluster head selection in wireless sensor networks. Neural Comput. & Applic. **32**, 7709–7723 (2020)
25. Xue, X., Ren, A.: An evolutionary algorithm based ontology alignment extracting technology. J. Netw. Intell. **2**(2), 205–212 (2017)
26. Pan, J.-S., Nguyen, T.-L.-P., Ngo, T.-G., Dao, T.-K., Nguyen, T.-T.-T., Nguyen, T.-T.: An optimizing cross-entropy thresholding for image segmentation based on improved cockroach colony optimization. J. Inf. Hiding Multim. Signal Process. **11**(4), 162–171 (2020)
27. Nguyen, T.-T., Nguyen, T.-D., Ngo, T.-G., Nguyen, V.-T.: An optimal thresholds for segmenting medical images using improved swarm algorithm. J. Inf. Hiding Multim. Signal Process. **13**(1), 12–21 (2022)
28. Liu, W., Hongbo, Y., Zhu, H., Fang, H.: Research on wireless sensor network localization based on an improved whale optimization algorithm. J. Internet Technol. **24**(1), 55–64 (2023)
29. Chang, W.-Y., Soma, P., Chen, H., Chang, H., Tsai, C.-W.: A hybrid firefly with dynamic multi-swarm particle swarm optimization for WSN deployment. J. Internet Technol. **24**(4), 825–836 (2023)
30. Pan, J.-S., Zhuang, J., Luo, H., Chu, S.-C.: Multi-group flower pollination algorithm based on novel communication strategies. J. Internet Technol. **22**(2), 257–269 (2021)

Chapter 11
Arduino-Based WMR with Solar Panel for Trajectory Tracking Control

Long-Chen Hung, Hsin-Tung Lan, Chun-Cheng Wei, Wei-Min Cheng, and Hsin-Tung Lan

Abstract This paper primarily investigates the use of Arduino as the core component in autonomous vehicles, complemented by the incorporation of solar panels to provide auxiliary power. The study aims to construct a wireless-controlled robotic vehicle with superior autonomy by utilizing four key components: the sensing circuit, motor circuit, Arduino microcontroller, and solar panel for backup power. The sensing circuit is employed to detect obstacles and facilitate obstacle avoidance through path correction, while the motor circuit governs the vehicle's motion. Solar panels serve as an alternative power source to enhance operational efficiency. The Arduino microcontroller serves as the central processing unit, orchestrating the synergy of all system components. Integration of hardware and software for autonomous navigation presents intriguing applications.

L.-C. Hung
Department of Electrical Engineering, Cheng Shiu University, Kaohsiung, Taiwan
e-mail: 0677@gcloud.csu.edu.tw

H.-T. Lan · C.-C. Wei (✉)
Department of Electronic Engineering, Lunghwa University of Science and Technology, Taoyuan, Taiwan
e-mail: ccwei@ccit.js.cn

W.-M. Cheng
School of Software and Big Data Changzhou College of Information Technology, Changzhou, China

H.-T. Lan
Department Information and Communication Engineering, Chaoyang University of Technology, Taichung, Taiwan

S.-P. Tseng et al. (eds.), *Advances in Intelligent Information Hiding and Multimedia Signal Processing, Volume 2*, Smart Innovation, Systems and Technologies 416, https://doi.org/10.1007/978-981-97-8760-9_11

11.1 Introduction

In recent years, the automotive industry has undergone a profound transformation driven by the global shift toward green energy economies and the urgent need to reduce greenhouse gas emissions [1–3]. Electric vehicles (EVs) have emerged as pivotal contributors to this transition, offering a promising solution to address environmental concerns. Nevertheless, the widespread adoption of EVs faces formidable challenges [4], with one of the most prominent being the protracted charging times. Even for short urban journeys, the extended charging durations of EVs can impose significant practical constraints [5, 6].

11.2 The System Description of WMR

This paper presents a novel approach to path tracking for a two-wheeled self-propelled vehicle, utilizing Arduino as the central driving unit and a solar panel as a backup power source. This vehicle boasts several distinctive features, including wireless control via Wi-Fi and Bluetooth modules, collision detection and avoidance capabilities, and a unique ability to navigate steep terrain, thereby enhancing stability and precision control [7, 8].

The intricately designed system offers precise movement control on level ground surfaces, along with robust data collection and processing capabilities. The design incorporates ultrasonic sensors, Bluetooth modules, Wi-Fi module control, and DC gear motors to govern the vehicle's speed and ensure collision avoidance by detecting obstacles in its path [9, 10].

The vehicle exhibits exceptional versatility, as it can operate in patrol and obstacle avoidance modes. It also offers path tracking functionality that can be remotely controlled from any location using Wi-Fi for long-range control and Bluetooth communication for short-range control, as shown in Fig. 11.1.

The WMR under consideration is positioned within a Cartesian coordinate system, and a fundamental assumption is made that precludes any lateral or sliding

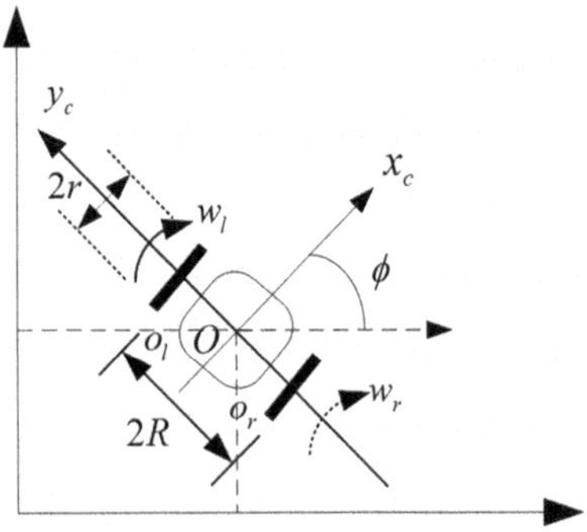

Fig. 11.1 The WMR Cartesian coordinate system

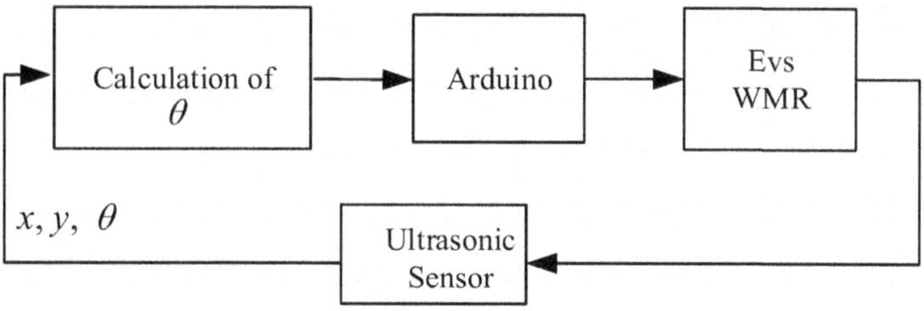

Fig. 11.2 The WMR for trajectory tracking control

movement. This foundational assumption consequently implies that the robot's motion primarily unfolds within the confines of the *x*-*y* plane, devoid of any lateral translations. Two pivotal parameters are introduced to characterize the robot's motion: The parameter '*v*' signifies the center speed of the WMR, serving as a representative measure of its linear velocity in the direction of travel [11].

The parameter *w* denotes the angular speed of the WMR, denoting the rate at which the robot undergoes rotational motion. The parameter *v* is the center speed, ω is the angular speed of the WMR, w_l is the speed of the left wheel, w_r is the speed of the right wheel, and *d* is the distance between the left wheel and right wheel. The velocity of the WMR on global coordinate is [11, 12].

$$\begin{bmatrix} x' \\ y' \\ \theta' \end{bmatrix} = \begin{bmatrix} \cos\theta \\ \sin\theta \\ 0 \end{bmatrix} \omega + \begin{bmatrix} 0 \\ 0 \\ 1 \end{bmatrix} \tag{11.1}$$

Within (11.1) the context of this analysis, the variable θ represents the heading angle, *x* corresponds to the *X*-axis coordinate denoting the center point of the entity, and *y* pertains to the *Y*-axis coordinate, also signifying the center point of said entity, as shown in Fig. 11.2. The computation of the distance traversed during motion can be derived as follows (11.2):

$$\begin{bmatrix} x_i\,(k+1) \\ y_i\,(k+1) \\ \theta_i\,(k+1) \end{bmatrix} = \begin{bmatrix} x_i(k) \\ y_i(k) \\ \theta_i(k) \end{bmatrix} + \begin{bmatrix} \cos\theta \\ \sin\theta \\ 0 \end{bmatrix} \begin{bmatrix} v \\ V \end{bmatrix} \Delta t \tag{11.2}$$

11.3 The Electrical System

Having done a lot of research and decisions based on the power systems of some of the most famous and best solar cars, we were able to design a better system that includes the best features of the best consumer cars. The electrical system of our

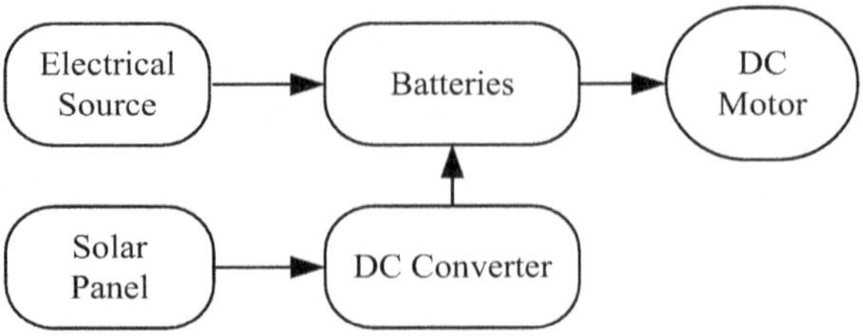

Fig. 11.3 The electric source with solar panels for DC motor

solar cars is the main source of energy. There are two main parts behind this, the first one deals with the basic energy source, which is solar energy, and the second one is the backup that is implemented to ensure that the car can be used no matter what the situation, even if it does't have sunlight [10, 13–15].

Electric vehicles are widely used due to their advantages such as easy maintenance, low operating costs, and environmental protection. Electrical energy stored in rechargeable batteries is the main source of power for electric motors in electric vehicles. The power consumed by photovoltaic cells is greatest at these locations. Short wavelength is determined by shorting the output channel and measuring the terminals. Photovoltaic cells are composed of semiconductor materials, the main components of which are crystals and thin films. In terms of cost and efficiency, most future thin materials are likely to outperform photovoltaic silicon cells, which can be divided into the following categories: crystalline, thin film, amorphous, multijunction, organic, and photochemical.

It captures photons from solar radiation and converts them into electrical energy. Connect directly to a converter, which converts the resulting voltage to direct current (DC). Electricity then flows to the motor so that the car can move, as shown in Fig. 11.3.

1. Solar Panels

The fundamental component within solar panels is the solar cell, also referred to as a photovoltaic cell. Its primary purpose is the conversion of solar energy into electrical energy. This conversion process relies on semiconductors, particularly employing a PN junction, which operates as a diode. Currently, the predominant type of solar cell in use is the crystalline silicon solar cell. These cells achieve a high level of efficiency through the incorporation of composite materials, allowing them to capture a broader spectrum of light compared to silicon alone. Each composite cell comprises three principal layers: germanium at the base, indium gallium arsenide or gallium arsenide in the middle, and indium gallium phosphide at the top.

2. Battery

Given the intermittent nature of solar energy, which is not available around the clock, the inclusion of batteries becomes paramount in ensuring a consistent power supply for the vehicle. The fundamental principle governing batteries revolves around chemical reactions: they comprise two electrode plates, positioned in close proximity to each other, and separated by a substance referred to as an electrolyte. Consequently, the flow of current instigates chemical reactions that involve the exchange of electrons, resulting in the generation of voltage. Batteries serve as reservoirs for electrical current by harnessing the pressure differential between the electrodes. In this particular context, lead-acid batteries are the preferred choice, albeit they incur a higher cost in comparison to nickel-cadmium batteries.

3. DC Motor

In the realm of solar-powered vehicles, the brushless DC motor emerges as the optimal choice, primarily attributed to its compatibility with the DC power generated by the previously elucidated power tracker. In comparison to their brushed DC counterparts, brushless DC motors offer a plethora of advantages. These encompass a superior torque-to-weight ratio, heightened torque efficiency, enhanced reliability, diminished acoustic emissions, an extended operational lifespan stemming from the absence of brushes and commutators prone to corrosion, the elimination of commutator-generated arcing, and an overall reduction in electromagnetic interference. Furthermore, owing to the absence of windings on the rotor, brushless DC motors remain unaffected by centrifugal forces. Additionally, the rotor's windings are housed within a supporting structure, enabling cooling through conduction mechanisms, obviating the need for internal airflow. When viewed from the perspective of a magnetic field generated by electrical current, the operational principles of a DC motor appear straightforward. Essentially, the motor operates within a closed-loop system where opposing forces are generated, culminating in rotational motion.

11.4 Experimental Results

This achievement bears immense significance, providing a robust foundation for the practical implementation of autonomous vehicle technology. Furthermore, the integration of solar panels as a supplementary power source has proven highly effective in extending the operational endurance of autonomous vehicles, substantially diminishing dependence on conventional batteries. This not only translates into cost savings but also contributes significantly to alleviating environmental pressures, thus endorsing the principles of sustainability. Our investigation has also underscored the remarkable potential of Arduino microcontrollers within the realm of autonomous vehicles, as shown in Fig. 11.4.

Their intrinsic flexibility and programmability render them the ideal choice for control systems, underscoring their pivotal role in achieving elevated levels of autonomy. Lastly, we wish to underscore the paramount importance and practical implications of this study. Our research presents an innovative approach that not only elevates the performance standards of autonomous vehicles but also propels the progression of autonomous vehicle technology, diversifying its application spectrum to encompass domains including, yet not limited to, autonomous driving vehicles,

Fig. 11.4 The Arduino-based electric vehicles with solar panel

drones, and robotics, as shown in Figs. 11.4, 11.5, 11.6 and 11.7. Moreover, the implications of our findings hold the potential to exert a positive influence on future research endeavors centered on energy efficiency and sustainability (Fig. 11.8).

In summation, the core objective of this study has been to seamlessly integrate both hardware and software components to enhance the autonomy and energy efficiency of autonomous vehicles. Therefore, we pave the way for promising prospects within the expansive realm of automation. We firmly believe that these discoveries will wield a profound and lasting impact on forthcoming technological developments and the promotion of environmental sustainability. We eagerly anticipate that our work will serve as an inspirational catalyst for further research in related fields, ultimately fostering the realization of more intelligent and efficient autonomous vehicle technologies.

11.5 Conclusion

In this study, we have conducted an in-depth exploration of the integration of Arduino and solar panels as a solution within the domain of autonomous vehicles, with the overarching goal of achieving enhanced autonomy and energy efficiency. Through the amalgamation of the sensing circuit, motor circuit, Arduino microcontroller, and solar panel as an auxiliary power source, several pivotal findings

Fig. 11.5 The Arduino-based WMR with Bluetooth module

Fig. 11.6 The Arduino-based WMR with Ultrasonic module

have emerged: First and foremost, we have successfully conceived and realized a wireless-controlled autonomous vehicle prototype, impeccably showcasing the harmonious synergy among sensing, motion control, and solar-powered backup systems.

Fig. 11.7 The Arduino-based WMR for obstacle avoidance

Fig. 11.8 The Arduino-based WMR for trajectory tracking control

Acknowledgments This paper is cooperated with Teaching Innovation Team of Changzhou College of Information Technology.

References

1. Sosa-Cervantes, C., Silva-Ortigoza, R., Márquez-Sánchez, C., Taud, H., Saldaña-González, G.: Trajectory tracking task in wheeled mobile robots: a review. In: International Conference on Mechatronics, Electronics and Automotive Engineering (2014)
2. Singh, S.A., Negi, A., Mudali, S.: Analysis of automatic sensing model of an autonomous vehicle. In: International Conference on Inventive Systems and Control (ICISC) (2017)
3. Barrit, D., Salih-Alj, Y.: Ralos car: solar powered car with a hybrid backup system. In: IEEE Symposium on Industrial Electronics and Applications (2012)
4. Anil, A., Jisha, V.: Trajectory tracking control of an autonomous vehicle using model predictive control and PID controller. In: International Conference on Control, Communication and Computing (ICCC) (2023)
5. Sadeghpour, D., Bauman, J.: Novel integrated electrical architecture for solar-charged electric vehicles. In: IEEE Transportation Electrification Conference & Expo (ITEC) (2022)
6. Benela, R., Jamuna, K.: Design of charging unit for electric vehicles using solar power. In: International Conference on Information Communication and Embedded Systems (ICICES) (2013)
7. Kaur, N., Bindal, R.: Modeling and simulation analysis of solar charging station for electric vehicle. In: 5th International Conference on Smart Systems and Inventive Technology (ICSSIT) (2023)
8. Prasad, B., Devi, T., Nookaraju, V., Sai, D., Kumar, P.: Modelling and simulation of solar powered hybrid electric vehicle. In: IEEE 8th International Conference for Convergence in Technology (I2CT) (2023)
9. Cutler, S.A., Schmalberger, B., Rivers, C.: An intelligent solar ecosystem with electric vehicles. In: IEEE International Electric Vehicle Conference (2012)
10. Mandal, S., Saw, S., Maji, S., Das, V., Ramakuri, S., Kumar, S.: Low cost arduino wifi bluetooth integrated path following robotic vehicle with wireless GUI remote control. In: International Conference on Information Communication and Embedded Systems (ICICES) (2016)
11. Chowdhury, M., Mamun, K., Rahman, M.: VModelling and simulation of power system of battery, solar and fuel cell powered hybrid electric vehiclee. In: 3rd International Conference on Electrical Engineering and Information Communication Technology (ICEEICT) (2016)
12. Kannan, M., William Mary, L., Priya, C., Manikandan, R.: Towards smart city through virtualized and computerized car parking system using arduino in the internet of things. In: International Conference on Computer Science, Engineering and Applications (ICCSEA) (2020)
13. Francisco, A., Lopes, N., Bento, L., Ferreira, C.: Arduino based open source electronic control unit for electric utility vehicles. In: International Scientific Conference Electronics (ET) (2020)
14. Krauss, R.: Combining raspberry pi and Arduino to form a low-cost, real-time autonomous vehicle platform. In: 2016 American Control Conference (ACC) (2016)
15. Kharade, J., Patil, A., Yadav, V., Kamble, B., Virbhadre, B.: Dual battery charger system for electric vehicle. In: Electronics and Sustainable Communication Systems (ICESC) (2020)

Chapter 12
AI-Powered Human Activity Recognition with Attention Networks

Muhammad Zia Ur Rehman, Murad Khan, and Anand Paul

Abstract Deep learning techniques effectively extract discriminative parameters from body-worn sensor data for human activity recognition (HAR). We propose an attention-based HAR model to enhance precision and interpretability. Our approach demonstrates superior recognition performance on benchmark datasets and provides attention maps, improving decision-making understanding. This innovation advances HAR and opens new research opportunities in AI-powered activity recognition, benefiting both academia and industry. The innovation presented in this paper not only contributes to the evolution of HAR methodologies but also marks a significant step forward in advancing the broader landscape of AI-powered activity recognition. By fostering a deeper understanding of intricate activity patterns through attention mechanisms, our approach not only benefits academia in refining state-of-the-art models but also holds promise for practical applications in industry. The insights gained from this research pave the way for new avenues of exploration and development in the realm of AI-driven human activity recognition.

12.1 Introduction

Human activity recognition (HAR) has become a vital tool in the domains of artificial intelligence and deep learning, with numerous applications in a variety of fields. In addition to being transformational, the ability of AI-powered systems to recognize and comprehend human behavior has the potential to completely alter consumer experiences, industrial procedures, and safety regulations. In this

M. Zia Ur Rehman (✉) · A. Paul
School of Computer Science and Engineering Kyungpook National University, Daegu, South Korea
e-mail: Ziaurrehmankhan921@gmail.com

M. Khan
Kuwait College of Science and Technology, Doha, Kuwait
e-mail: m.khan@kcst.edu.kw

S.-P. Tseng et al. (eds.), *Advances in Intelligent Information Hiding and Multimedia Signal Processing, Volume 2*, Smart Innovation, Systems and Technologies 416, https://doi.org/10.1007/978-981-97-8760-9_12

Fig. 12.1 Human activity recognition

context, our research explores the HAR space, concentrating on leveraging attention networks to improve precision and interpretability. The empirical validation of our proposed model unfolds on the WISDM dataset, showcasing its prowess through superior recognition performance when benchmarked against established metrics. Moreover, our model furnishes interpretable attention maps that illuminate the decision-making processes, providing a holistic view of its discernment. This paper aims to contribute significantly to the ongoing discourse in HAR research, with a specific emphasis on the application of deep learning techniques to the WISDM dataset. Beyond academic implications, our work holds promise for practical applications in diverse industries, unlocking new possibilities in AI-powered activity recognition (Fig. 12.1).

Nowadays, with the increasing usage of smartphones, smart watches, smart outfits, and other similar gadgets, several embedded sensor devices (such as accelerometers, gyroscopes, and cameras in smartphones) are widely used, providing effective study instruments for HAR investigations [1–6]. An extensive range of daily actions, including standing, walking, sitting, running, and lying, are recognized by several HAR systems using accelerometers. In order to identify and prevent falls in older persons in a smart environment, the scientists looked through accelerometer data to locate repetitive activities [7–12]. The majority of the aforementioned systems employ several accelerometers that are fixed at various points on the human body. Due to the numerous sensors attached to the human body and cable connections, this technique, however, has an impact on how individuals live their daily lives.

Central to our investigation is the conceptualization and implementation of an attention-based HAR model, a sophisticated framework designed to discern and comprehend the nuanced intricacies of human activities. The incorporation of attention mechanisms elevates our model beyond conventional approaches, endowing it with the ability to selectively focus on critical aspects of the sensor data, enhancing precision and interpretability. This model, a manifestation of the synergy between cutting-edge deep learning and the wealth of information embedded in the WISDM dataset, seeks to redefine the benchmarks for accuracy and comprehension in HAR systems.

Recent years have seen a notable improvement in the extraction of discriminative features from raw sensor data for human activity identification (HAR) employing body-worn sensors. Our study utilizes attention mechanisms to address the imminent need to develop more precise and comprehensible HAR models. We present a novel self-attention-based HAR architecture that exhibits outstanding recognition performance across a range of tasks on validated datasets. Our approach differs from others in that it can provide attention maps, which sheds information on the model's decision-making process and makes it suitable to a wide range of industrial domains.

12.2 Related Work

Human activity recognition (HAR) is a pivotal approach that plays a crucial role in applications ranging from healthcare monitoring to smart environments. This research explores the integration of attention-based neural network into HAR systems, aiming to overcome challenges associated with centralized models, particularly in terms of data privacy and scalability. In addition, mobile phones' built-in portability and computing power have made them an essential part of our everyday existence [13]. On the other hand, contrasting elements, complex backgrounds, lighting variations, occlusion of target objects, and variations in illumination provide performance issues for vision-based activity detection algorithms [14]. Consequently, because smartphone sensor data has so many advantages over other sensor modalities, it has become the main focus of human activity recognition research [15].

For traditional machine learning methods to transform raw sensor data into a feature representation, significant feature engineering and domain expertise are required. This representation is then utilized by classifiers like support vector machines [16] or CNN [17], could recognize actions. Convolutional and recurrent neural networks are two examples of deep learning algorithms that are particularly good at automatically extracting features and classifying them. They have shown encouraging outcomes in a variety of sensor-based scenarios for human activity detection [18].

Accurately detecting or recognizing human actions is still a difficulty, even with the many automated human activity recognition (HAR) techniques put forth in the literature. Among the main challenges is the effective representation of sensory input. Traditional methods for surface machine learning, such as support vector machine, decision trees, and Naïve Bayes [19], are contingent deeply on the excellence of hand-crafted topographies, such as statistical [13] and also frequency domain structures. Moreover, although hand-crafted features are capable of accurately representing simple human motor actions, they are frequently unable to discriminatively capture complicated behaviors. Feature engineering techniques are very domain-specific and difficult to effectively generalize to novel and complicated data sources and experimental setups [20].

12.3 Methodology

By addressing existing issues with the literary works and in seeking to overcome the limitations inherent in current literature, our research endeavors to establish a robust and accessible model for the identification of a diverse array of human activities. This includes, but is not limited to, walking, running, jogging, and more, leveraging accelerometer data collected from a mobile device strategically worn around the waist. The goal is to enhance the comprehensibility and efficacy of human activity recognition (HAR) models, contributing to advancements in the field., using accelerometer data obtained from a mobile device worn around the waist. Dataset preparation is the first step in the HAR process. Training, validation, and test sets are created from the WISDM (wireless sensor data mining) dataset, which contains sensor data collected from smartphones during a variety of activities. Next, normalizing sensor readings, turning them into fixed-length sequences, and encoding activity labels are all included in data preprocessing. The design of an attention-based neural network forms the basis of the process. With the use of this attention mechanism, the model can highlight crucial segments of the sensor data sequence, improving accuracy. In order to avoid overfitting, performance on the validation set is monitored as the network learns from the training set while applying the proper loss functions and optimization algorithms. The model's accuracy, precision, recall, and F1-score are evaluated on the test set. The model is optimized by hyperparameter tuning, and the visualization of attention weights offers insights into how the model makes decisions. After applying post-processing for noise reduction and approving the findings, the model can be used for batch or real-time inference on new sensor data. This thorough methodology guarantees the successful detection of human actions on the WISDM dataset using attention-based neural networks.

12.3.1 Dataset Description

In the field of human activity recognition research, the wireless sensor data mining (WISDM) accelerometer dataset demonstrated in Fig. 12.2. is a key resource.

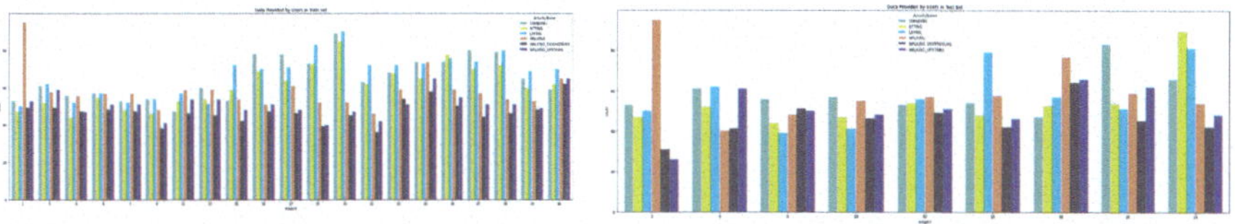

(a) Data Provided by User in Train Set (b) Data Provided by User in Test Set

Fig. 12.2 Dataset description (**a**) Data provided by user in train set. (**b**) Data provided by user in test set

This dataset, which was gathered from 51 people, shows a range of six different activities, including walking, jogging, moving upstairs, downstairs, sitting, and standing. Participants were told to carry a smartphone with accelerometer and gyroscope sensors in their pockets while performing the prescribed exercises in order to collect data. The dataset, which primarily focuses on accelerometer data but may also include gyroscope data, is adaptable for a variety of analytical uses. For those delving into the intricacies of activity recognition or data analysis, the official WISDM Lab website serves as a valuable gateway, providing convenient access to this extensive and diverse dataset.

Researchers and data analysts can leverage this resource to advance their studies, experiment with novel methodologies, and contribute to the ongoing discourse in the field of human activity recognition. The WISDM accelerometer dataset, with its rich and varied data, plays a crucial role in fostering advancements and insights in understanding human movements in diverse contexts.

12.3.2 Model Training

During the model training phase, we feed the attention-based neural network that was created with the preprocessed sensor data sequences from the training dataset. The model gains the ability to recognize temporal connections and patterns in the data during this phase. For multi-class classification, we use categorical cross-entropy as an appropriate loss function to quantify the discrepancy between the predicted activity labels and the actual activity labels. In order to reduce this loss and promote convergence, optimization methods Adam alter the network's weights and biases. Performance is optimized by fine-tuning hyperparameters like learning rate, batch size, and network architectural specifics. In order to avoid overfitting and decide when to terminate training using strategies like early stopping, we closely monitored the performance measures of the validation dataset throughout training. These variables include accuracy and loss. This iterative approach guarantees that the model generalizes effectively and develops the ability to precisely identify human actions based on the sensor data of the WISDM dataset. The dataset used

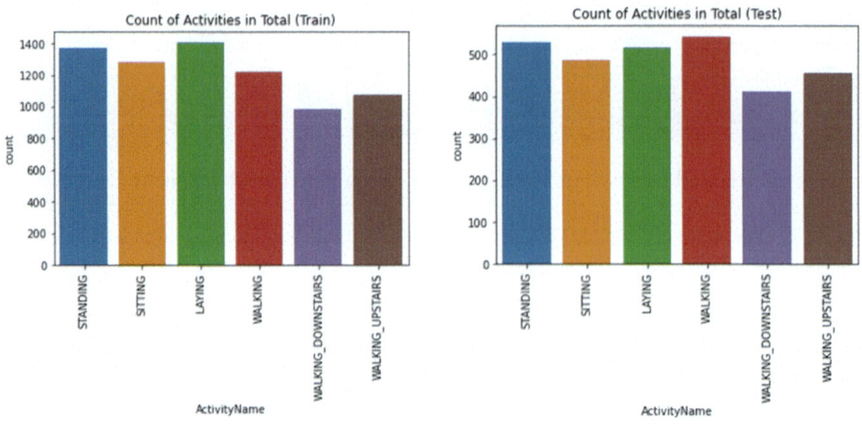

(a) Count of Activitties in Train Set (b) Count of Activitties in Test Set

Fig. 12.3 Dataset description (**a**) Count of activities in train set. (**b**) Count of activities in test set

for model training is shown in Fig. 12.3. The methodology provided below in algorithmic form makes it simple to find out more about the proposed framework.

Algorithm 1: AI-Powered Human Activity Recognition with Attention Networks

Data: Preprocessed data: $X_{\mathtt{train}}$, $y_{\mathtt{train}}$, $X_{\mathtt{val}}$, $y_{\mathtt{val}}$.
Result: Trained model: $M_{\mathtt{trained}}$.
Initialize neural network M; Set hyperparameters: α, B, E; $e \leftarrow 0$;
while $e < E$ **do,**
  Divide $X_{\mathtt{train}}$, $y_{\mathtt{train}}$ into mini-batches of size B;
  foreach *mini-batch* **do,**
    Perform forward pass through M to get predictions;
    Compute loss between predictions and true labels;
    Backpropagate gradients and update model weights;
  Evaluate M on $X_{\mathtt{val}}$, $y_{\mathtt{val}}$;
  if *Validation metrics do not improve* **then,**
    Stop training and save M_{trained}; $e \leftarrow e + 1$;
$M\mathtt{trained} \leftarrow M$;
The trained model is ready for inference;

12.4 Results and Discussion

Utilizing machine learning approaches to HAR techniques is becoming increasingly popular. Our study exploits and expands attention-based neural networks to simulate data with various interactions more effectively. As it can be seen from the table, the

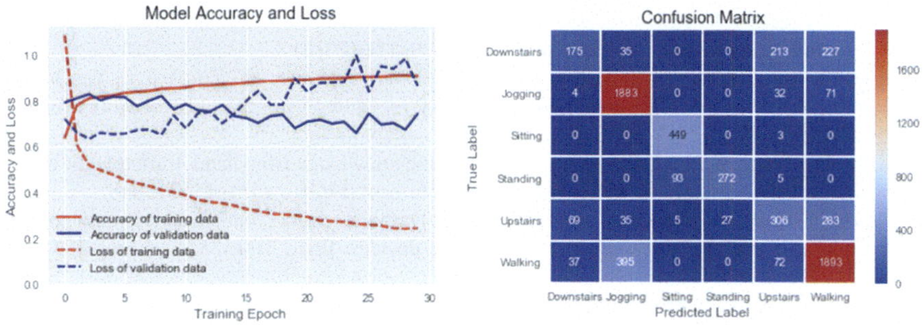

(a) Model Accuracy (b) Confusion Matrix

Fig. 12.4 Proposed model results. (**a**) Model Accuracy. (**b**) Confusion Matrix

Table 12.1 Results table

Models	Accuracy	Precision	Recall	F1
Attention-based NN	97.75	96.39	95.56	97.05

proposed algorithm achieved 97.75% accuracy rate with 96.39% precision, 95.56% recall, and 97.05% F1 score respectively. The algorithm's ability to accurately identify and classify human activities, as evidenced by its high precision, recall, and F1 score, signifies a robust performance across a spectrum of activities. This not only enhances the overall reliability of HAR systems but also positions attention-based neural networks as effective solutions in diverse real-world scenarios (Fig. 12.4 and Table 12.1).

12.5 Conclusion

In conclusion, this study has shed light on how attention-based neural networks are used for human activity recognition (HAR) on the basis of the WISDM dataset. We have proven the effectiveness of attention processes in improving HAR accuracy by careful preprocessing, modeling, and training. The findings support attention-based neural networks' promise as a formidable HAR tool, with wide-ranging implications for industries including healthcare and IoT. This work sets the path for further developments in activity recognition and associated applications while also adding to the body of knowledge in machine learning.

References

1. Yang, Y., Hou, C., Lang, Y., Guan, D., Huang, D., Xu, J.: Open-set human activity recognition based on micro-doppler signatures. Pattern Recogn. **85**, 60–69 (2019)

2. Saini, R., Kumar, P., Roy, P.P., Dogra, D.P.: A novel framework of continuous human-activity recognition using kinect. Neurocomputing. **311**, 99–111 (2018)
3. Khan, M.U.S., Abbas, A., Ali, M., Jawad, M., Khan, S.U., Li, K., Zomaya, A.Y.: On the correlation of sensor location and human activity recognition in body area networks (bans). IEEE Syst. J. **12**(1), 82–91 (2016)
4. Chen, Z., Zhang, L., Cao, Z., Guo, J.: Distilling the knowledge from hand-crafted features for human activity recognition. IEEE Trans. Industr. Inform. **14**(10), 4334–4342 (2018)
5. Khalifa, S., Lan, G., Hassan, M., Seneviratne, A., Das, S.K.: Harke: human activity recognition from kinetic energy harvesting data in wearable devices. IEEE Trans. Mob. Comput. **17**(6), 1353–1368 (2017)
6. Lv, M., Chen, L., Chen, T., Chen, G.: Bi-view semi-supervised learning based semantic human activity recognition using accelerometers. IEEE Trans. Mob. Comput. **17**(9), 1991–2001 (2018)
7. Anguita, D., Ghio, A., Oneto, L., Parra, X., Reyes-Ortiz, J.L., et al.: A public domain dataset for human activity recognition using smartphones. In: ESANN, vol. 3, p. 3 (2013)
8. Ghio, A., Oneto, L., et al.: Byte the bullet: learning on real-world computing architectures. In: ESANN (2014)
9. Noor, M.H.M., Salcic, Z., Kevin, I., Wang, K.: Adaptive sliding window segmentation for physical activity recognition using a single tri-axial accelerometer. Pervasive Mob. Comput. **38**, 41–59 (2017)
10. Lee, Y.-S., Cho, S.-B.: Activity recognition with android phone using mixture-of-experts co-trained with labeled and unlabeled data. Neurocomputing. **126**, 106–115 (2014)
11. Chen, Z., Zhu, Q., Soh, Y.C., Zhang, L.: Robust human activity recognition using smartphone sensors via ct-pca and online svm. IEEE Trans. Ind. Inf. **13**(6), 3070–3080 (2017)
12. Cao, L., Wang, Y., Zhang, B., Jin, Q., Vasilakos, A.V.: Gchar: an efficient group-based context—aware human activity recognition on smartphone. J Parallel Distrib Comput. **118**, 67–80 (2018)
13. Essa, E., Abdelmaksoud, I.R.: Temporal-channel convolution with self-attention network for human activity recognition using wearable sensors. Knowl.-Based Syst. **278**, 110867 (2023)
14. Raj, T., Nisar, T., Abbas, M., Priyadarshini, R., Naz, S., Tiwari, U.: Qualitative analysis of techniques for device-free human activity recognition. Revue d'Intelligence Artificielle. **37**(3) (2023)
15. Al-Hassani, R.T., Atilla, D.C.: Human activity detection using smart wearable sensing devices with feed forward neural networks and pso. Appl. Sci. **13**(6), 3716 (2023)
16. Anguita, D., Ghio, A., Oneto, L., Parra, X., Reyes-Ortiz, J.L.: Human activity recognition on smartphones using a multiclass hardware-friendly support vector machine. In: Ambient Assisted Living and Home Care: 4th International Workshop, IWAAL 2012, Vitoria-Gasteiz, Spain, December 3–5, 2012. Proceedings 4, pp. 216–223. Springer (2012)
17. Jiang, W., Yin, Z.: Human activity recognition using wearable sensors by deep convolutional neural networks. In: Proceedings of the 23rd ACM International Conference on Multimedia, pp. 1307–1310 (2015)
18. Chen, K., Zhang, D., Yao, L., Guo, B., Yu, Z., Liu, Y.: Deep learning for sensor-based human activity recognition: overview, challenges, and opportunities. ACM Comput. Surv. (CSUR). **54**(4), 1–40 (2021)
19. Jatoba, L.C., Grossmann, U., Kunze, C., Ottenbacher, J., Stork, W.: Contextaware mobile health monitoring: evaluation of different pattern recognition methods for classification of physical activity. In: 30th Annual International Conference of the IEEE Engineering in Medicine and Biology Society, pp. 5250–5253. IEEE (2008)
20. Rokni, S.A., Nourollahi, M., Ghasemzadeh, H.: Personalized human activity recognition using convolutional neural networks. In: Proceedings of the AAAI Conference on Artificial Intelligence, vol. 32, (2018)

Chapter 13
A System for Automatic Generation of Code and Simulation Data

Yao Lu, Xin Guo, Zhenyu Zhou, and Junfeng Chen

Abstract At present, most developers still use duplicate entity code for development and write large amounts of simulation data to assist in testing. It leads to long project completion time and low quality of work. This paper studies the code and simulation data generation and develops a web system for users. Based on Springboot technology, the system converged different data types into TableSchema objects and passed them to the generator to generate the required object code and simulation data. During this process, the FreeMarker template engine was used to generate Java and front-end code. Multiple types of simulation data were generated using different tools or custom methods, such as using Generex to generate regular expression type data. By automatically generating code and simulation data, the system improves the efficiency of user's program development.

13.1 Introduction

The traditional software development model has many problems, one of which is repeated coding, which is becoming increasingly prominent, especially in the field of web application development. With the expansion of web application scale, the amount of manual code that developers need to write is also increasing exponentially. How to quickly develop web applications that meet the needs of enterprises has become particularly crucial [1]. Because most web applications are information systems, developers typically must write a large amount of similar functional code to adapt to different database table structures. Meanwhile, due to the different personal styles of developers, the consistency and quality of the code cannot be fully guaranteed. This not only affects the quality of the final program, but also makes maintenance work more complex.

Y. Lu · X. Guo (✉) · Z. Zhou · J. Chen
School of Information Science and Technology, Sanda University, Shanghai, China
e-mail: guoxin@sandau.edu.cn

S.-P. Tseng et al. (eds.), *Advances in Intelligent Information Hiding and Multimedia Signal Processing, Volume 2*, Smart Innovation, Systems and Technologies 416, https://doi.org/10.1007/978-981-97-8760-9_13

Another major issue is the longer testing time required. Before the software is officially launched, the development team must conduct a series of tests, including functional testing and automation testing, to ensure the smooth operation of the system.

Furthermore, as the testing workload increases, testers need a large amount of simulation data to meet system requirements. This results in testers having to spend a lot of time writing these simulation data and engaging in monotonous and repetitive work. In summary, more and more software developers and testers need a way to quickly generate code and simulate data to support their work. Especially for development work that requires handling many repetitive tasks, automated code generation and simulation data generation tools will significantly improve development efficiency and reduce error rates. Code generation and simulation data generation tools can automatically generate code and data that meet specific requirements, helping developers quickly build and test applications. These tools are usually able to generate a large amount of code and data within a few minutes, thereby shortening the software development cycle and improving development efficiency.

13.2 Related Work

Code generation technology aims to utilize automated tools to generate the source code required for applications, to improve development efficiency. With the continuous advancement of its development process, researchers have adopted various technical methods to achieve code generation. For example, the S-function is used to design a CAN message sending and receiving model that can import DBC files. Ruiteng Y et al. proposed a CAN message unpacking and packaging method based on Simulink automatic code generation technology to address the issues of large data volume and low efficiency in traditional software development for automotive controller CAN communication messages [2]. Jessica E L et.al comprehensively reviewed the latest methods for natural language generation in Java code, using RNN models or Transformer models to solve this problem, and made significant progress in this field [3]. Rathinaraja Jeyaraj et.al in order to solve the challenge of reliable decision-making in big data, is to use Hadoop MapReduce for cost-effective processing. A multilevel per node combinator for MapReduce jobs is proposed, which reduces the intermediate data and improves the completion time by 32.4% compared with the traditional scheduler using the default combinatory [4].

With the introduction of model-driven architecture technology, automatic code generation technology in China has begun to develop rapidly. For example, Xianmin Wu proposed a method of automatically generating corresponding code using a decoding layer of long-term and short-term memory networks combined with algorithms [5]. Tianyu Liu proposed the use of automatic code generation technology to automatically convert models into control code. Based on the open source graphical simulation platform Si Ros/Xcos, code generation for the hyperblock model is implemented [6]. These tools can automatically generate code and data, and can be customized and extended through configuration files and templates, helping developers complete system development, reducing the difficulty and complexity of coding, and shortening the development time of the system.

In addition, many small and medium-sized enterprises and open source institutions have also performed well in automatic code generation, developing small code generators that can support the conversion of models (templates) to various programming language codes. For example, dynamic soft.NET code generator is a C # automatic code generator based on a relational database. CodeMatic generates code based on object-oriented thinking and a three-layer architecture design; The Code Generator generation tool is an engine that generates code based on Velocity templates [7].

Data generation technology, which many researchers have explored and studied from different perspectives, is an important research direction in the field of software testing. Model-based test data generation technology has also received widespread attention and research. This method models test objects and generates test data based on the model. Xuedi Gao, etc. have proposed an automated test data generation method based on an improved genetic algorithm, aiming to improve the efficiency and quality of software testing [8].

The potential advantage of the system developed in this paper over existing tools is that the TableSchema constructor is used to synthesize existing popular generators and manage them in a unified way. It makes the system more flexible and expandable, higher maintainability, easy to understand, and improve development efficiency. By abstracting table structures of different data types into TableSchema objects, the system is able to do code generation more efficiently. Developers can implement complex code and simulated data generation requirements by simply defining TableSchema. This can help shorten development cycles and improve development efficiency, especially when code and data with different table structures need to be generated frequently. It can also reduce the occurrence of human logic errors.

13.3 Requirement Analysis of the System

13.3.1 Functional Requirement Analysis

Based on the investigation and analysis of user needs, the code and simulation data generation system needs to have the following functional modules:

Login Module. This module includes login and registration functions. Users can enter their username, password, and verification code to log in, or register by filling in their personal information.

System Tool Module. This module provides functions such as form construction and one click generation, which includes Java code, front-end code, JSON code,

SQL statements, and simulated data. Users should have the ability to copy and download the results, and be able to save and call the forms constructed by users.

Business Management Module. Users can save the results after generating code or simulating data, or make calls in the forms.

User Management Module. It is used to manage user data, including user registration, login, and permission data management.

13.3.2 Business Process Analysis

Before creating the system, we need to sort out the whole business process of the project. The business process of the system is shown in Fig. 13.1.

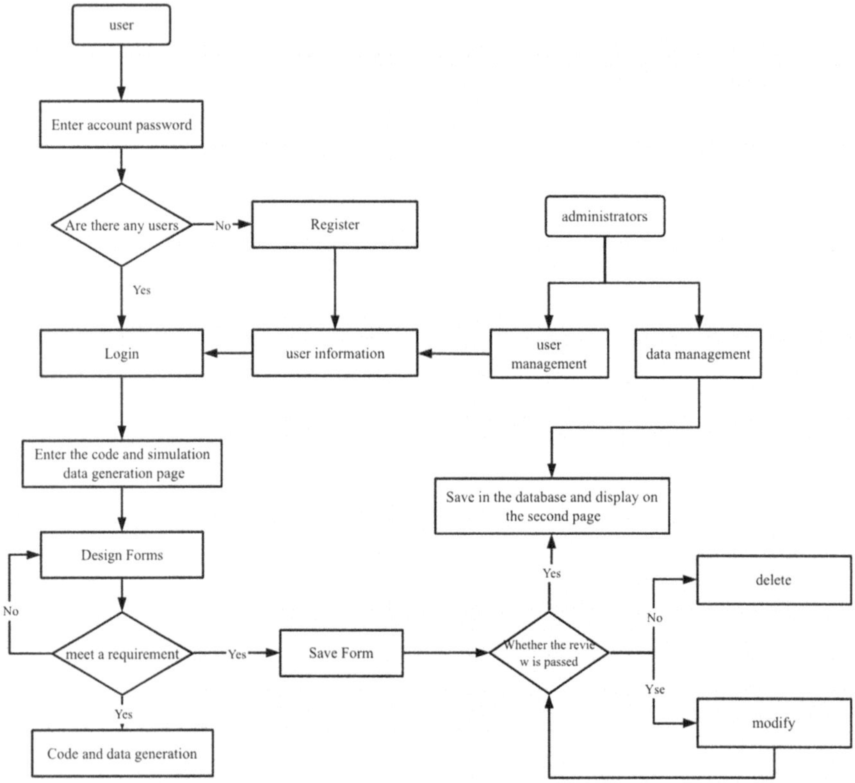

Fig. 13.1 The automatic generation system aims to improve the development efficiency of programmers. Users can customize form settings and generate code and simulation data after logging in. In addition, users can save the designed form content, which can be stored in the database and presented on the page after being reviewed by the administrator

13.4 Design of System

13.4.1 Design of Overall System Functional

Design system functional modules based on requirement analysis, including user login registration module, code and data generation module, and system management module, are shown in Fig. 13.2.

13.4.2 Design of Code Generation Function Java Code Generation

Java entity classes and object code encapsulation. JavaEntityGenerateDTO uses reflection technology to generate DTO code for Java entity classes based on user provided entity class information, and encapsulates the generated code into a class. This class can be used directly or its methods can be called to obtain DTO code for Java entity classes.

Freemarker template engine. JavaEntityGenerateDTO and JavaObject GenerateDTO use Freemarker as the template engine. They pass the encapsulated Java entity class code and Java object class code to the template engine for code construction by loading predefined template files. In Freemarker templates, syntax such as variables, loops, and conditions can be used to generate corresponding Java code based on the attribute information of entity classes. The overall workflow of Java code generation is shown in Fig. 13.3.

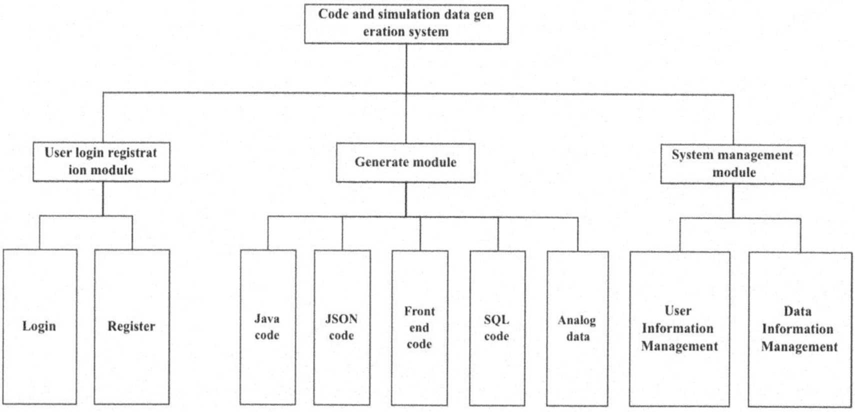

Fig. 13.2 The figure shows the overall modular composition of the system

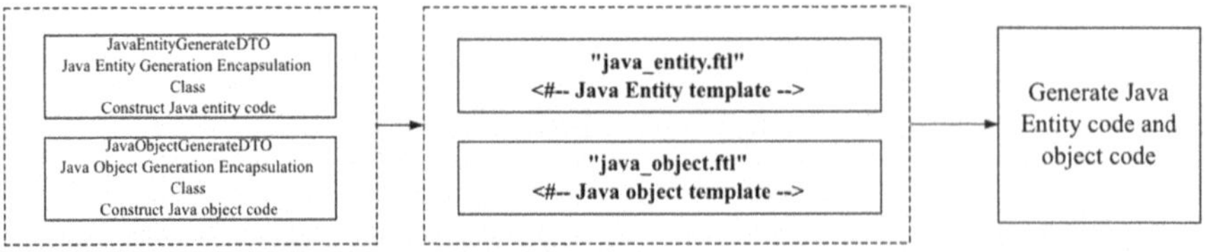

Fig. 13.3 JavaEntityGenerateDTO and JavaObjectGenerateDTO output generated Java code to the specified front-end window for developers to use. Developers can copy the generated Java code and integrate it into their systems

Fig. 13.4 The Gson object converts the mapping list into a JSON string, serving as the method's return value

JSON Code Generation A static method called buildJson is provided, which takes a mapping list as input to represent data rows, where each mapping corresponds to a data row and the key corresponds to a column name. In this method, use the GsonBuilder class to create a Gson object. Finally, we use the toJson method of the Gson object to serialize the data into a JSON string. The toJson method converts Java objects into JSON strings for output. The overall workflow of JSON code generation is shown in Fig. 13.4.

Frontend Code Generation

Encapsulate front-end Typescript code. TypescriptTypeGenerateDTO uses reflection technology to generate DTO code for Java entity classes based on user provided entity class information, and encapsulates the generated code into a class. This class can be used directly or its methods can be called to obtain DTO code for Java entity classes.

Freemarker template engine. TypescriptTypeGenerateDTO uses Freemarker as the template engine, passing encapsulated front-end Typescript types to the template engine for code construction by loading predefined template files. In Freemarker templates, syntax such as variables, loops, and conditions can be used to generate corresponding front-end code based on the attribute information of entity classes. The overall workflow of front-end code generation is shown in Fig. 13.5.

SQL Code Generation In the SQL dialect factory pattern, define an abstract factory interface and then define a specific Factory class. These specific factory classes are responsible for creating specific SQL dialect instances so that applications can use them to interact with the database. In this way, the overall workflow of SQL code generation is shown in Fig. 13.6.

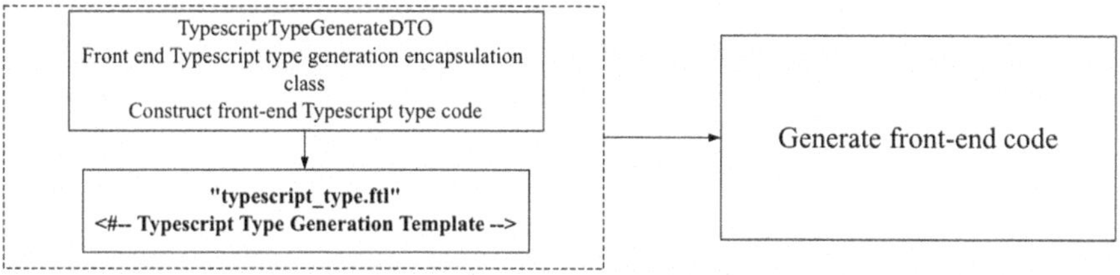

Fig. 13.5 TypescriptTypeGenerateDTO outputs the generated front-end code to the specified front-end window for developers to use. Developers can copy the generated front-end code into the developed system

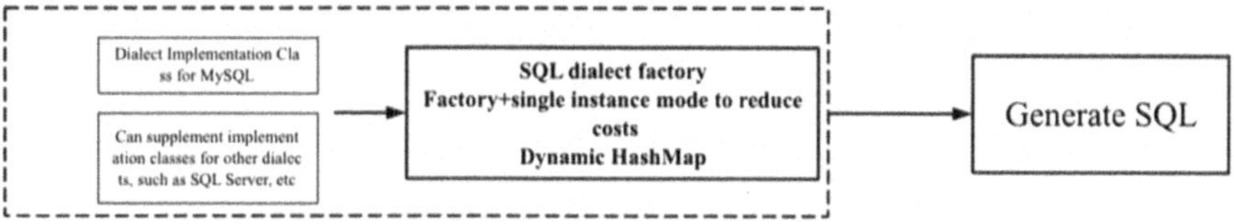

Fig. 13.6 Switch between different databases without changing the application code

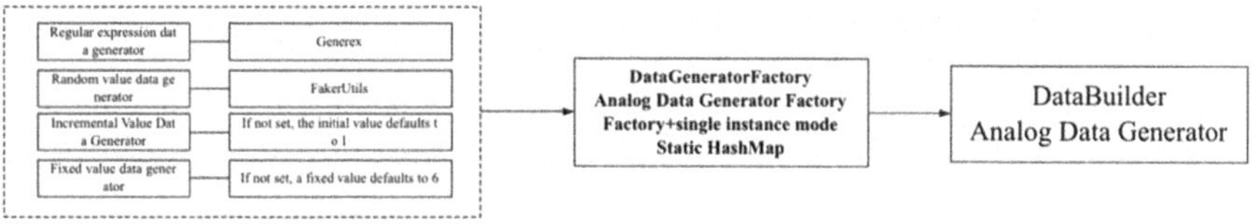

Fig. 13.7 The figure shows that regular expressions use Generex and random numbers use FakerUtils. Incremental data and fixed values are written separately and other methods are called

13.4.3 Design of Simulation Data Generation Function

Define a specific class for each data type and write all these classes to the data factory using a unified interface. In the data factory, create a static HashMap so that each generator corresponds to a key, and then package them as a whole so that the control layer can call them as a whole. The workflow of simulating data generation is shown in Fig. 13.7.

13.4.4 Design of Database

The conceptual structure design of the database makes the relationships between data clearly visible, highlighting various internal structures within the database. The E-R model is shown in Fig. 13.8.

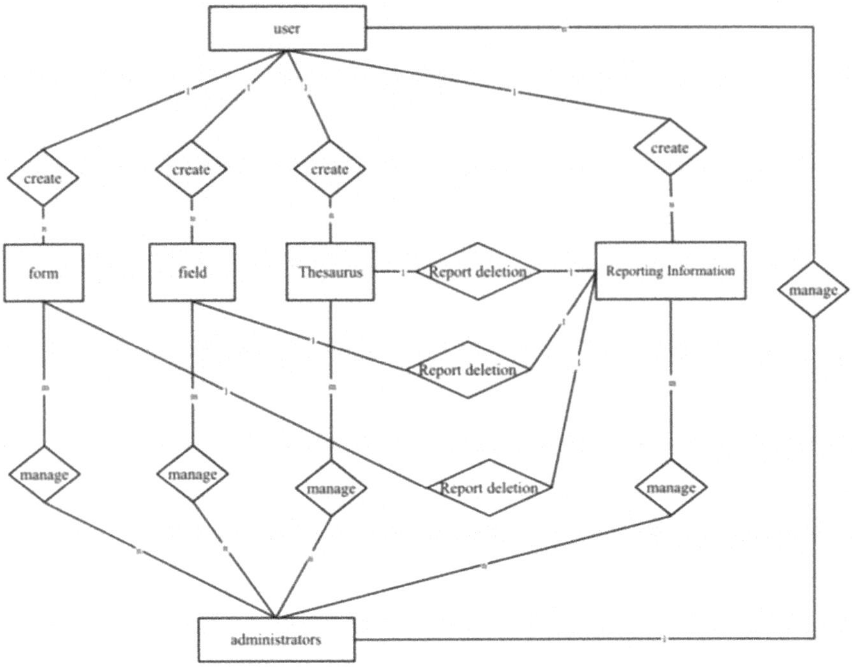

Fig. 13.8 Use the E-R model to describe the conceptual structure of a database

13.5 Implementation of System

13.5.1 Implementation of Java Code and Fronted Code Generation

Since the Java code and the front-end code are designed on the same principle, the implementation sections are presented together. Both use the FreeMarker template engine to generate code based on table schemas and supplied data. The setConfiguration method is used to inject the Configuration object. The final effect of all the Java generated parts is shown in Fig. 13.9.

13.5.2 Implementation of JSON Code Generation

The buildJson method uses the Google Gson library to convert input data into a JSON string. The GsonBuilder class is used to create a Gson instance that supports beautiful printing, and it formats the JSON output into a format more suitable for human reading. This method returns a JSON array string containing all the JSON

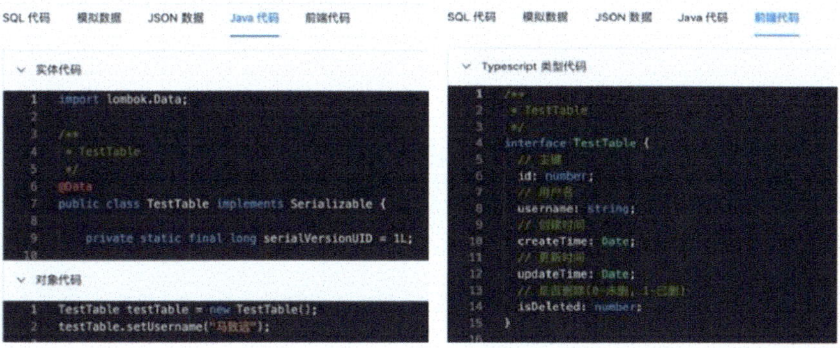

Fig. 13.9 The generated Java code and front-end code have fields corresponding to the columns in the table whose values are populated by values from the first row of data

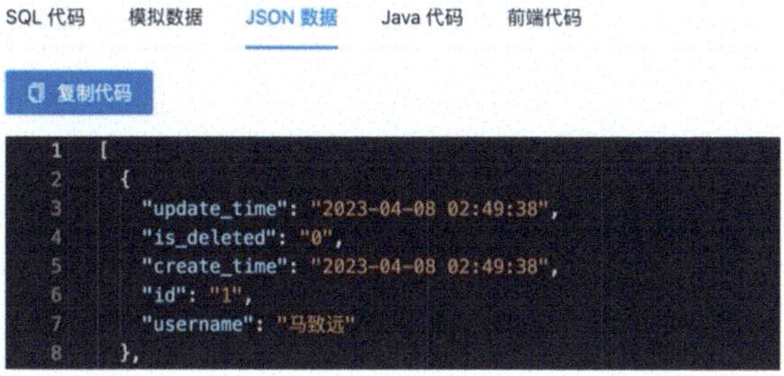

Fig. 13.10 The input data is a list of mappings, where each mapping represents a JSON object. The key in each map is the field name and the value is the corresponding field value

objects in the input list. If the input list is empty, this method returns an empty JSON array [9]. The final effect of all the JSON generated parts is shown in Fig. 13.10.

13.5.3 Implementation of SQL Code Generation

For SQL data insertion, the process starts with a template string for table names, column names, and values. The table name is from the TableSchema object, formatted using the wrapTableName method of the sqlDialect object. The field list is filtered for column names and values. For each data row, the method generates an INSERT statement by filling in the template string. The result is a string representation containing all SQL statements for inserting data into the database.

13.5.4 Implementation of Simulation Data Generation

The factory plus singleton pattern is applied to these 5 simulated data types in the DataGeneratorFactory class. The singleton pattern ensures that a certain class has at most one instance and provides this instance to the entire system, that is, the class needs to provide a global method to access the unique instance. This class is called the singleton class. In the simulation data generation function, five random simulation data types can be chosen, including fixed value, increasing value, random value, and regular expression.

13.6 Test of System

13.6.1 Test Result

The system testing includes three parts: user module, code module, and data generation module. Six test cases are written in the user module. The code and data generation module compiled 22 test cases. 28 test cases passed the functional test, 0 test cases failed. Performance tests show that the response time of the system is less than 500 milliseconds. Security tests have also been successfully passed, with 0 potential vulnerabilities found.

13.6.2 Specificity and Reusability of System

The implementation of the project is to construct different generators into the same type by single encapsulation of factory pattern and singleton pattern into the Tableschema constructor. So whenever the designer or user wants to add additional functionality or output code. No need to change the existing code. Simply wrapping a single new generator into the TableSchema hub can help shorten development cycles and improve efficiency, especially when code and data with different table structures need to be generated frequently. It can also reduce the occurrence of human logic errors. This reflects code reusability.

The code output for this system also has a strong reuse. The user is not required to know the professional code knowledge, only needs to modify the field according to the idea. The project also provides the copy and download of the output results.

13.7 Conclusion

The development of a code and simulation data generation system aims to improve the development and testing efficiency of technical personnel and simplify the complexity of data processing. It provides developers and learners with fast access to simulated data needed for testing, as well as ways to learn SQL language and related codes. The system has a variety of code generation and simulation data generation systems. If used on a large scale, the information in the library can be classified according to different categories, forming a dictionary system, will be more convenient for related industry personnel to learn. In this way, users can more easily find the information they need, such as classification according to data types, functions, syntax, etc., to meet the needs of users at different levels and in different fields. At the same time, such a classification can also allow learners to better understand and master the knowledge points of SQL language, so as to more effectively use SQL skills for data processing and analysis. In addition, with the continuous application of SQLHome, the storage of analog data will become more meaningful, and the analog data will be stored in ks, which is more convenient for storage and query. It can also improve and enrich the content and functions of the system through user feedback and communication, so as to better meet the needs of users and improve the learning effect.

Acknowledgments This work was sponsored by Shanghai Municipal Education Commission under the contract Z90004.23.001 (Professional Master's Degree Authorized School Training Project), the contract C2024057 (Educational Science Research Project), and the Key Undergraduate Education Reform Project (Intelligent Teaching Quality Evaluation Model and System Development Based on Multimodal Data Fusion).

References

1. Zhou, B.: Design and Implementation of Java Backend Code Generator Based on SSM Framework. Dalian University of Technology (2020)
2. Yuan, R., Xiao, C., Liu, M., et al.: CAN message unpack and pack based on simulink automatic code generation technology. J. Phys. Conf. Ser. **2232**(1), 012009 (2022)
3. Jessica, E.L., Sanoussi, M.A.Y., Mehdi, E.C., et al.: A comprehensive review of state-of- the-art methods for Java code generation from natural language text. J. Nat. Lang. Process. **2023**(3), 100013 (2023)
4. Jeyaraj, R., Ananthanarayana, V.S., Paul, A.: Fine-grained data-locality aware MapReduce job scheduler in a virtualized environment. J. Ambient Intell. Human Comput. **11**, 4261–4272 (2020)
5. Wu, X.: Research on Code Generation Technology Based on Deep Learning. East China Normal University (2022)
6. Liu, T.: Research on Automatic Code Generation Based on Open Source Simulation Platform. North University of China (2021)
7. Zhang, Y., et al.: Analysis and comparison of mainstream code generation methods. Comput. Modernizat. **11**, 103–108 (2015)

8. Gao, X., et al.: Research on automatic generation of test data based on improved genetic algorithm. Comput. Sci. **44**(03), 209–214 (2017)
9. Song, C., et al.: Analysis of the programming contribution of automatic generation and completion of application code. Inf. Technol. **46**(12), 29–46 (2022)

Chapter 14
Telehealth 2.0: A Novel Framework for Real-Time Liveliness Detection and Mood Analysis

Barathi Subramanian, Rakhmonov Akhrorjon Akhmadjon Ugliz, Bahar Amirian Varnousefaderani, and Jeonghong Kim

Abstract Telemedicine is revolutionizing healthcare delivery, especially in remote areas and situations where face-to-face consultations are infeasible. However, as the adoption of digital health solutions accelerates, it brings forth challenges related to the authenticity of remote sessions. A significant threat in recent years has been spoofing attacks, such as photo impersonations or video replays, which may threaten the authenticity of remote interactions. It is even more difficult for healthcare professionals to gauge a patient's emotional well-being in the absence of physical signals. This can lead them to miss the subtle signs of distress or anxiety. To address these multifaceted issues, this paper introduces a novel framework for real-time liveliness detection and mood analysis. By merging the capabilities of the YOLO (you only look once) v8 model enhanced with spatial attention mechanism (SAM) and frequency analysis (FA) using fast Fourier transform (FFT) integrated with a deep convolutional neural network (CNN). Preliminary findings, showcased through real-time emotion tracking and comprehensive emotion distribution charts, highlight the unique ability of our system to accurately measure emotions in real time. This not only offers profound insights into a patient's mental state but also paves the way for proactive healthcare interventions in telemedicine.

14.1 Introduction

Healthcare has always been dynamic, evolving in response to scientific advancements and societal needs. In particular, telemedicine demand was precipitated by the COVID-19 pandemic, cementing its position as a vital component of outpatient healthcare delivery [1]. The American Telemedicine Association (ATA) encapsulates this evolution, defining telemedicine as the utilization of electronic

---

B. Subramanian · R. A. A. Ugliz · B. A. Varnousefaderani · J. Kim (✉)
Kyungpook National University, Daegu, South Korea
e-mail: jhk@knu.ac.kr

S.-P. Tseng et al. (eds.), *Advances in Intelligent Information Hiding and Multimedia Signal Processing, Volume 2*, Smart Innovation, Systems and Technologies 416, https://doi.org/10.1007/978-981-97-8760-9_14

means to convey medical information, seamlessly connecting patients and providers separated by distances [2, 3]. A central element of telemedicine's origin lies in its promise of bridging geographical divides, and democratizing healthcare by reaching people in remote areas or who cannot travel for in-person consultations. There is no doubt that it has the potential to transform healthcare delivery, but it has also brought with it a new set of challenges.

In recent years, advanced spoofing techniques, from simple photo impersonations to elaborate video replays, have emerged as important threats, posing considerable risks to the trust and authenticity that are at the heart of patient-healthcare provider relationships [4, 5]. Such breaches bring with them not only a hazard in terms of technical security but also the possibility of eroding the bedrock of trust upon which the framework of healthcare is built. Additionally, while telemedicine has made it easier to deliver healthcare remotely, it has also displaced medical professionals from non-verbal indications that patients frequently display in the absence of verbal communication. When a healthcare interaction relies solely on digital technology, these distinctions can be overlooked, potentially resulting in incomplete or misdirected care, as they are vital for the complete care of the individual [6].

To address these multifaceted challenges, this paper aims to present an efficient telemedicine solution that can be tailored to the burgeoning telemedicine ecosystem. A novel framework has been developed for the YOLO (you only look once) v8 model, which merges the capabilities of the YOLO model with spatial attention mechanism (SAM) and frequency analysis (FA) using the fast Fourier transform (FFT), combined with a deep convolutional neural network (CNN). Through this dual-pronged approach, not only is the authentication of a session ensured, but also the emotional distance inherent in digital interactions is bridged.

14.2 Related Works

Over the past few years, telemedicine has made significant strides as it has demonstrated its ability to redefine healthcare delivery, particularly in situations where face-to-face consultations are not feasible. In an increasingly globalized healthcare landscape, telemedicine offers a promising solution, increasing access to medical consultations, regardless of geographical limitations [4, 5]. There are, however, many challenges associated with the use of this digital approach, including ensuring genuine interaction between the patient and the doctor and decoding non-verbal signs from the patient [6].

The need to validate the authenticity of patients increased as digital consultations became more common. According to Chingovska et al., facial recognition systems are vulnerable to threats including photo impersonation, videos, and 3D masks [7]. Early countermeasures, such as the facial texture analyses discussed by Maatta et al., provided foundational solutions [8]. Nevertheless, advanced defensive strategies were required due to the continuously evolving threat landscape. With the recent development of deep learning methods, facial dynamics can be effectively leveraged for detecting liveness [9].

Telemedicine faces additional challenges beyond authenticity, including the interpretation of a patient's emotional state. Traditional consultation provides healthcare professionals with a wealth of insights into patient's emotional and psychological well-being [10]. In the context of digital healthcare, emotion recognition provides a powerful solution. There was an early emphasis on the value of emotion-aware systems in healthcare put forth by Picard et al. [11]. With the advancement of technology, deep learning models, such as the work of Khorrami et al., have shown impressive fluency in analyzing facial expressions with remarkable precision [12, 13].

Even though advances have been made in both liveness detection and emotion recognition within telemedicine, gaps remain, particularly when seamlessly integrating these systems into a complete solution.

14.3 Proposed Methodology

Our proposed framework for telehealth 2.0 comprised of three pivotal stages: Liveliness detection, emotion recognition, and real-time system for emotion alerts. An overview of our proposed end-to-end emotion aware decision support framework is shown in Fig. 14.1. The following sections explain each of the above in detail.

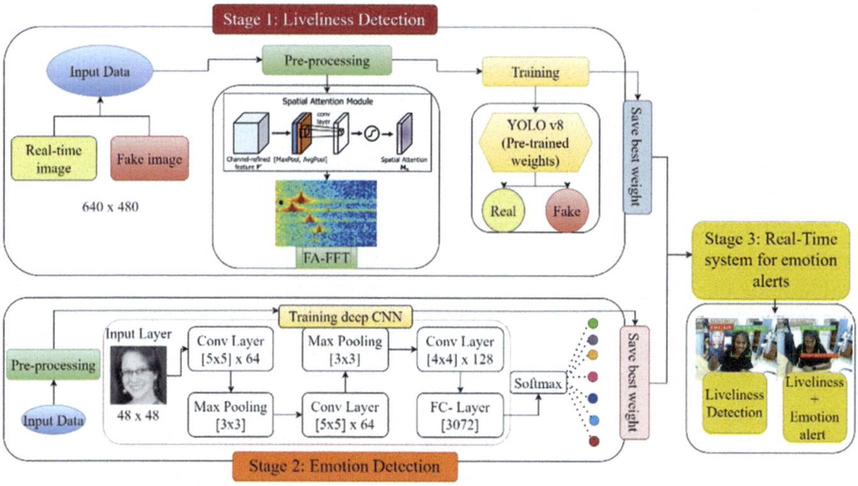

Fig. 14.1 Overall framework for the proposed system

14.3.1 Stage 1: Liveness Detection

An essential component of any secure telehealth system is the assurance that the participant is truly present. By using webcam, we captured video frames in real time, providing a continuous data stream that can be used for further analysis. This contains real-time collection (real person sitting in front of the camera) and collection from various sources(mobile phone images, online videos streams, and pictures from books) to make the system robust. We collected 2960 images for each class (fake and real).

To focus on areas of the image that contain significant information, particularly those enhancing the clarity and details of faces, we implemented a SAM before processing each frame. The spatial attention module directs the model's focus to salient regions within the image, thus prioritizing facial features more likely to exhibit signs of liveliness, such as eye movements or subtle expressions. The spatial attention mechanism can be mathematically represented as follows:

$$Attention\,(x, y) = \sigma\,(W\cdot \tanh\,(U\cdot f\,(x, y))) \tag{14.1}$$

where *Attention(x,y)* denotes the attention weight at spatial location *(x,y)*, σ is the sigmoid activation function, W and U are learnable weights of the model, and $f(x, y)$ is the feature map. The output of this module is a modified feature map with enhanced regions interest. The input image has dimensions $H \times W \times C$ (height, width, channels), the output of the SAM maintains these dimensions but with modified pixel intensities.

Following spatial attention, we perform FA using FFT on the enhanced frames. This process transforms the spatial features into the frequency domain, highlighting temporal changes indicative of liveliness:

$$X(f) = \int_{-\infty}^{\infty} x(t)\cdot e^{-j2\pi ft}dt \tag{14.2}$$

The FFT output for each frame is a complex-valued frequency domain representation of shape H $\times$ W $\times$ C. We then take the magnitude of this complex output, maintaining the spatial dimensions but transforming the feature space to emphasize frequency characteristics.

The outputs of both the SAM and FFT are concatenated along the channel dimension, resulting in a combined feature map of dimensions H $\times$ W $\times$ 2C. This concatenated feature map effectively combines spatial attention-enhanced features with frequency domain information, providing a comprehensive representation of each frame for liveliness detection. This enriched feature map is then fed into the YOLOv8 model. YOLOv8, with its convolutional layers, is adept at processing such multi-dimensional data. The model is trained to classify each frame into 'real' or 'fake' categories based on these concatenated features. We employ the pre-trained weights of YOLOv8 as initialization and train the model for 50 epochs with a batch size of 32, using an 80–20 split for training and testing. The final output of the YOLOv8 model is the classification result for each input frame, determined by a SoftMax classifier.

During training, the YOLOv8 model learns to interpret both the spatially focused and frequency-highlighted aspects of the input, enhancing its ability to discern real human presence from fake representations. This combination of attention and frequency analysis ensures that the model is not only trained on superficial image features but also on underlying temporal patterns that are characteristic of live entities. The best weights from this training are saved for integration into the later stages of our telehealth framework.

In enhancing the robustness of liveness detection, our model incorporates dynamic temporal analysis to discern genuine human presence. By training on a diverse dataset that includes both genuine and artificial representations of individuals, our system can effectively recognize and mitigate spoofing attempts, a critical security measure in telehealth applications. This proactive approach not only refines the system's accuracy but also instills trust in the telehealth process, ensuring that interactions are legitimate and secure. In the future development of this stage, we plan to explore the integration of thermal imaging and 3D depth sensing to further augment the system's discernment capabilities, thus advancing the security framework for telehealth platforms.

14.3.2 Stage 2: Emotion Recognition

For emotion recognition various datasets are available for research. For our model we used FERC-2013 [14] dataset that is comprised of 32,000 low-resolution images. This consists of 7 various emotions like happy, sad, surprise, neutral, angry, fear, and disgust as shown in Fig. 14.2. We pre-process the data and pass it to a deep CNN [15] as shown in Fig. 14.1 for training.

This initial stage involves preparing the input data for processing. The sample input data shows a greyscale image of a face, which suggests that the pre-processing step likely includes normalization or standardization of pixel values, resizing, and possibly face detection and alignment. The processed image is then fed into an input layer of the neural network, which has a dimensionality of 48×48 pixels. This implies that the image is resized to a 48×48 pixel square before it's passed into the network. The network includes multiple convolutional layers (Conv Layer), where each layer has a set of filters that perform convolutional operations. The first convolutional layer has 64 filters of size 5×5. After the convolution operation, a max pooling layer with a 3×3 window is applied to reduce the spatial dimensions of the output. The architecture shows a repeating pattern of a convolutional layer followed by a max pooling layer. After the first convolutional layer and pooling, there's another convolutional layer with 64 filters of size 5×5, followed by another max pooling layer. This is followed by a third convolutional layer with 128 filters of size 4×4. After the convolutional layers, the architecture has a fully connected

Fig. 14.2 Training samples from FER-2013

layer with 3072 units. Fully connected layers are typically used to flatten the output of the convolutional layers and combine features for classification. At the end of the network, a SoftMax function is used. The SoftMax function is often employed in the final layer of a neural network-based classifier to represent the probabilities of the outcomes, in this case, the different emotions that the network is trying to detect. The network was trained for 20 epochs with 80% images for training and 20% for testing. Finally, after training the model, the output layer classifies the 7 emotions. After training, the best-performing weights (parameters) of the network are saved. This implies that during the training process, the model is evaluated, and the configuration that gives the best performance is preserved for stage 3.

14.3.3 Stage 3: Real-Time System for Emotion Alert

To create a real-time system for telehealth applications, we integrated the two trained models from stage 1 and stage 2 by utilizing the best weight saved for each model as mentioned above. A webcam on the computer is used to capture video frames, which are then analyzed in real time, so immediate feedback is given. Through the enhanced training methodology, each frame is processed by the initialized YOLO model to detect the presence of faces. In order to categorize a face as 'live' or 'fake,' the system uses the processed face region as input. Using the original video frame, this decision is visualized, along with its confidence level, to provide users with a better understanding. Green indicates real faces, while red indicates fake ones as shown in Fig. 14.3. Once the detector finds the input feed as real, it will display the current mood of the person with high-risk alert for negative emotions as shown in Fig. 14.3b.

The conclusion of our framework is the real-time emotion alert system, which serves as a sentinel for patient engagement and well-being. This stage represents the integration of technological proficiency with clinical sensitivity. The immediacy of feedback ensures that any signs of distress are not overlooked, and appropriate responses can be initiated without delay. The potential to extend this system includes adaptive learning capabilities where the model personalizes alerts based on patient history and preferences, thereby refining the telehealth service to be more patient-centric.

14.4 Experimental Setting and Results

In this section, we present the results of our experiments demonstrating the effectiveness of the system in detecting real-time emotions and detecting liveness. We ran the simulation using NVIDIA GeForce RTX 3080 Ti with Python 3.9 version with RAM memory-32GB. In the conducted experiments we achieved mean average precision score (mAP)@0.5 as 99% for liveliness detection and 95% test accuracy for emotion recognition model.

14.4.1 Liveliness Detection Analysis

As part of the initial analysis, the goal was to distinguish between genuine ('Real') and spoofed ('Fake') video feeds. The difference between the real and fake data samples can be clearly seen in Fig. 14.4. The pie chart representation reveals that 89.7% of feeds were genuine, while 10.3% were detected as spoofing. By making this distinction, the framework emphasizes its ability to differentiate and potentially combat spoofing.

14.4.2 Real-Time Emotion Tracking

We evaluated the capability of our system to detect emotions in real time. Figure 14.5 illustrates how detected emotions have evolved over a specific period of time.

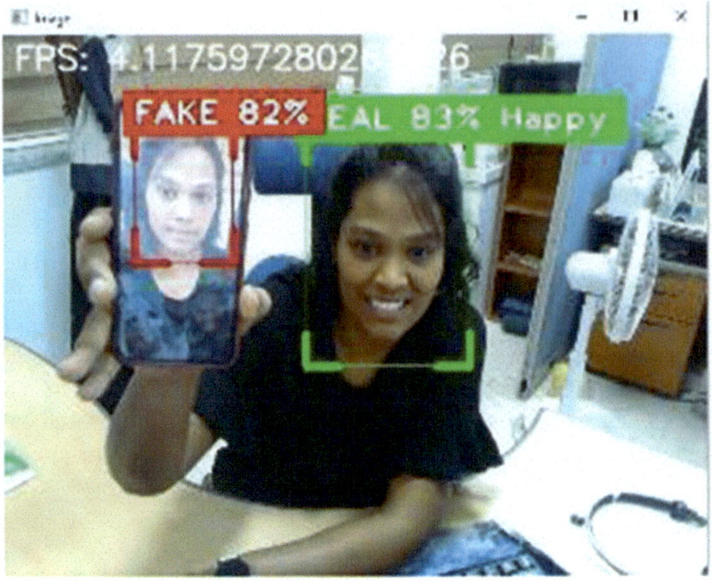

(a)

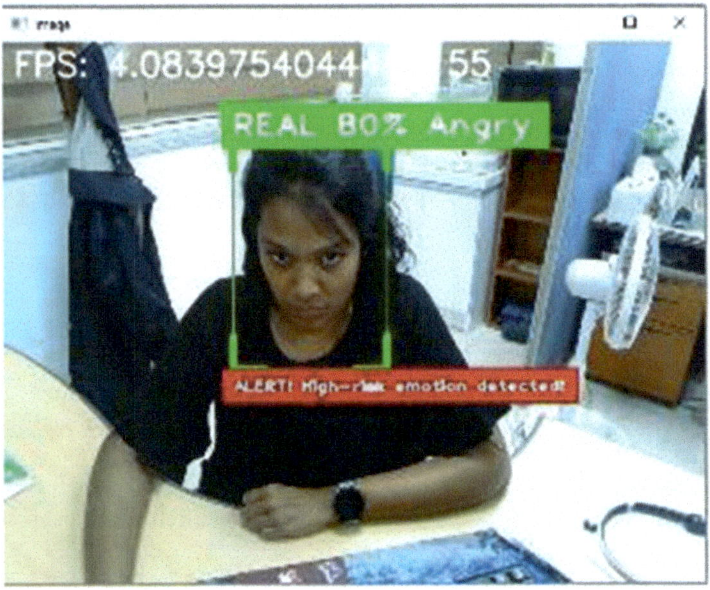

(b)

Fig. 14.3 Visualization of real-time system's output. (**a**) Liveliness detection and (**b**) liveliness detection with high-risk alert for negative emotions

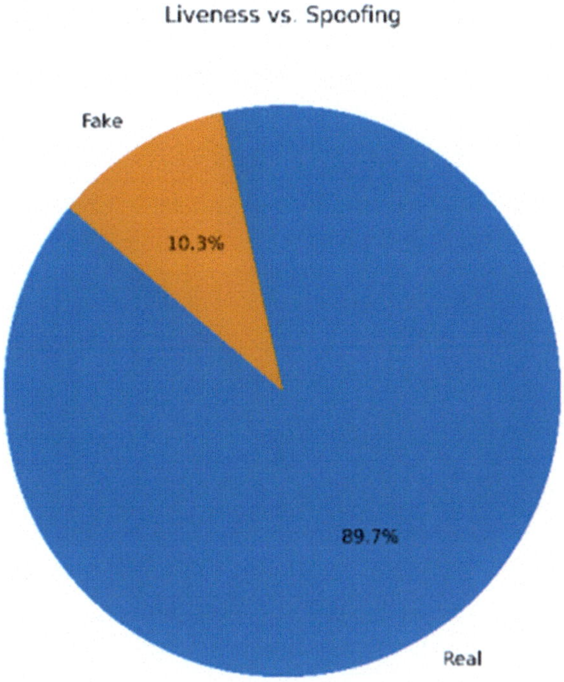

Fig. 14.4 Pie chart representation of real and fake video feed in real time

Line graphs represent emotions identified at particular timestamps as points on the graph. The x-axis indicates the progression of time, and the y-axis indicates the emotions that have been detected. As a result of this temporal emotion tracking, we are able to understand the emotional dynamics of a telehealth session, allowing us to offer timely interventions or support to those in need.

14.4.3 Emotion Distribution Analysis

In order to understand the emotional landscape during telehealth consultations, an emotion distribution analysis was conducted. As shown in Fig. 14.6, the frequency of each detected emotion during the experiment is represented by a bar chart. This graph shows the distinct emotions on the x-axis, and the counts on the y-axis. In view of this distribution, healthcare professionals can tailor their approach based on the predominant emotions they encounter during telehealth sessions.

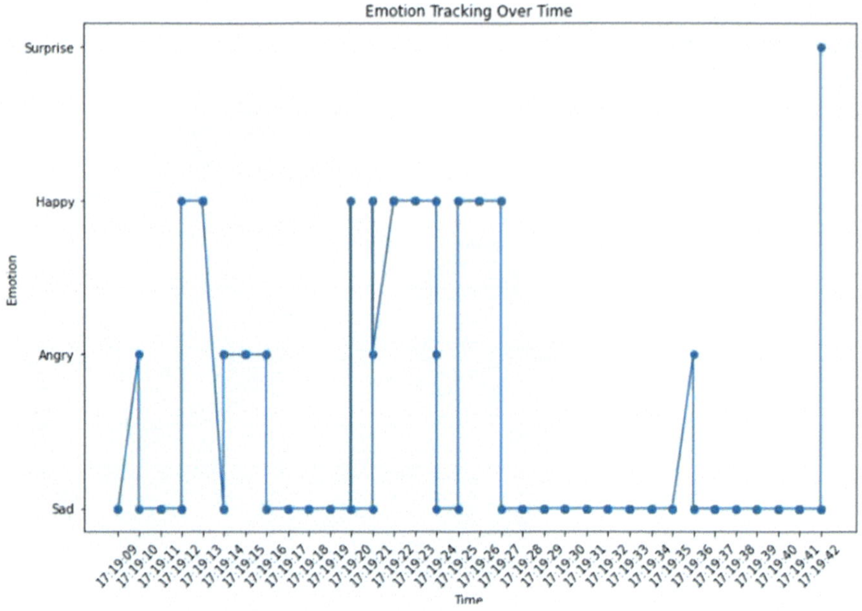

Fig. 14.5 Real-time emotion tracking over a period of time

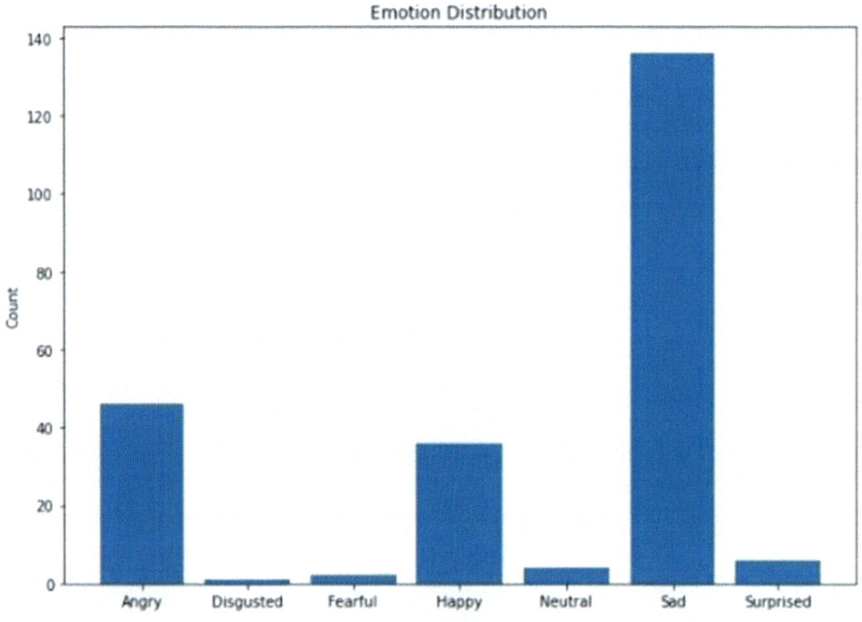

Fig. 14.6 Emotion distribution bar chart for detected emotions in real time

14.5 Conclusion

Despite innovations like telemedicine, authenticating sessions and recognizing emotions have proven challenging. With this in mind, our study developed an integrated framework capable of incorporating the YOLO v8 model enhanced with SAM and FFT integrated with a deep CNN. Results from our study demonstrate that the system is able to detect emotions in real time and provides insights into the well-being of patients. As a result of this innovation, telehealth consultations are not only enriched, but they are also kept secure, and patient-centered.

Acknowledgments This study was supported by the BK21 FOUR project (AI-driven Convergence Software Education Research Program) funded by the Ministry of Education, School of Computer Science and Engineering, Kyungpook National University, Korea (4199990214394).

References

1. Arshad Ali, S., et al.: Global interest in telehealth during COVID-19 pandemic: An analysis of Google Trends™. Cureus. **12**, e10487 (2020). https://doi.org/10.7759/cureus.10487
2. Hui, D.S., et al.: The continuing 2019-nCoV epidemic threat of novel coronaviruses to global health – the latest 2019 novel coronavirus outbreak in Wuhan, China. Int. J. Infect. Dis. **91**, 264–266 (2020). https://doi.org/10.1016/j.ijid.2020.01.009
3. American Telemedicine Association.: Telemedicine glossary [EB/OL] (2019)
4. Haimi, M.: The tragic paradoxical effect of telemedicine on healthcare disparities – a time for redemption: a narrative review. BMC Med. Inform. Decis. Mak. **4**, 1–10 (2023). https://doi.org/10.1186/s12911-023-02194-4
5. Giansanti, D.: Ten years of telehealth and digital healthcare: where are we? Healthcare (Switzerland). **11**(6) (2023). https://doi.org/10.3390/healthcare11060875
6. Schueller, S.M.: Grand challenges in human factors and digital health. Front. Digit. Heal. **3**(May), 1–4 (2021). https://doi.org/10.3389/fdgth.2021.635112
7. Chingovska, I., Anjos, A., Marcel, S.: On the effectiveness of local binary patterns in face anti-spoofing. In: 2012 BIOSIG-Proceedings of the International Conference of Biometrics Special Interest Group (BIOSIG), pp. 1–7. IEEE (2012)
8. Määttä, J., Hadid, A., Pietikäinen, M.: Face spoofing detection from single images using texture and local shape analysis. IET Biom. **1**(1), 3–10 (2012). https://doi.org/10.1049/iet-bmt.2011.0009
9. Yu, Z., Qin, Y., Li, X., Member, S., Zhao, C.: Deep learning for face anti-spoofing: a survey. IEEE Trans. Pattern Anal. Mach. Intell. **45**(5), 5609–5631 (2023). https://doi.org/10.1109/TPAMI.2022.3215850
10. Ekole, A.: Nonverbal communication in clinician-patient interaction and influence on healthcare outcome. Am. J. Trade Policy. **9**(1), 43–50 (2022). https://doi.org/10.18034/ajtp.v9i1.619
11. Picard, R.W.: Affective Computing. MIT Press (2000)
12. Le Paine, T., Huang, T.S.: Do deep neural networks learn facial action units when doing expression recognition? In: Proceedings of the IEEE International Conference on Computer Vision Workshops, pp. 19–27 (2015)
13. Subramanian, B., Kim, J., Maray, M., Paul, A.: Digital twin model: a real-time emotion recognition system for personalized healthcare. IEEE Access. **10**, 81155–81165 (2022). https://doi.org/10.1109/ACCESS.2022.3193941
14. Kaggle.: Challenges in representation learning: facial expression recognition challenge (2013)
15. Gudi, A.: Recognizing semantic features in faces using deep learning (2015)

Chapter 15
Moving Object Detection Method Based on Improved Visual Background Extractor with Weighted Local Binary Patterns

Zheng Xuan Tao, Xiao Peng Wang, and Liang Long

Abstract In response to the issues of ghost regions in traditional ViBe (visual background extractor) algorithm during motion target detection, as well as the potential for false positives or misses in the presence of dynamic interference in complex video frames, an improved ViBe algorithm is proposed. Firstly, improve the initial background model using the five-frame differencing method, and then employ the weighted local binary patterns (WLBP) algorithm to extract texture features from the frame difference images for the suppression of ghost regions. Secondly, during the foreground detection process, adaptively calculate the radius threshold using a dynamic adjustment of the update rate based on the scene's complexity, enhancing resistance to interference. Lastly, perform morphological operations and connectivity analysis on the extracted foreground objects, fill in the detected foreground object gaps to address deficiencies in motion object detection. The experimental results demonstrate that the improved algorithm, as compared to the traditional ViBe algorithm, enhances detection accuracy across various test scenarios such as subway stations, boats, highways, and railways. Specifically, the accuracy improved by 11.6, 20, 11.3, and 10.8% in these respective scenarios. The recall rate also showed significant improvements, with increases of 18.8, 22.3, 11.8, and 12.3%. The comprehensive evaluation metric, F-M value, increased by 0.134, 0.196, 0.111, and 0.117 in these scenarios. The proposed method effectively reduces the occurrence of ghost regions, achieving lower false positive and false negative rates, and significantly improves accuracy compared to the traditional ViBe algorithm.

Z. X. Tao (✉) · X. P. Wang · L. Long
College of Electronic and Information Engineering, Lanzhou Jiaotong University, Lanzhou, China
e-mail: wangxiaopeng@mail.lzjtu.cn

S.-P. Tseng et al. (eds.), *Advances in Intelligent Information Hiding and Multimedia Signal Processing, Volume 2*, Smart Innovation, Systems and Technologies 416, https://doi.org/10.1007/978-981-97-8760-9_15

15.1 Introduction

Motion object detection refers to the identification, localization, and tracking of moving objects in images or videos. It involves three steps: motion detection, object localization, and target tracking. This technology finds wide applications in fields such as video surveillance and intelligent transportation systems. At present, motion object detection methods based on video sequences primarily include optical flow, interframe differencing, and background differencing, with background differencing being the most commonly used technique for motion object detection. Optical flow is a technique that computes the motion values of pixels between adjacent frames of images. It determines the motion speed and direction by analyzing differences in image brightness values, thereby enabling the detection and tracking of motion objects. Common background differencing methods include GMM, CodeBook, and ViBe [1]. GMM exhibits significant adaptability to complex conditions but still requires the selection of a background model based on model data. The ViBe algorithm is a pixel-level background modeling method. It begins by sampling the neighborhood around each pixel and initializing the pixel's background model based on the sampling results [2]. During the detection process, the ViBe algorithm calculates the difference between the current frame image and the background model. It classifies pixels using a specific thresholding strategy, assigning them as either foreground or background. For pixels labeled as background, their background models are updated. For pixels labeled as foreground, direct foreground detection is performed. This updated strategy allows for adaptation to scene changes and accurate extraction of motion objects. The ViBe algorithm, due to its use of simple random sampling, is designed to create a background model that can effectively detect and track moving objects by exploiting changes between the background model and the current frame information. This approach offers real-time performance and relatively low computational complexity [3]. However, it also has its limitations. When the initial background model significantly differs from the actual scene or contains partial moving objects, these moving objects may be incorrectly classified as part of the background, leading to the appearance of ghost regions. Moreover, in cases where the background experiences subtle changes or drifts, such as lighting variations or swaying tree leaves, applying fixed thresholds may result in erroneous foreground labeling. In such situations, the ViBe algorithm may struggle to accurately track these changes, leading to false positives or false negatives.

To address the issue of ghost regions in the ViBe algorithm, literature [4] calculates the offset of the video and then modifies the ViBe modeling approach based on this offset. This alteration changes the range of ViBe pixel sets and consequently affects the sensitivity of the model. However, it results in subsequent foreground targets being detected with missing parts. In literature [5], a method is proposed that uses multiple consecutive frames instead of a single frame to initialize and reduce ghosting. It also employs adaptive thresholds instead of a global fixed threshold. Nonetheless, this approach is relatively complex and may not be suitable for widespread application. In literature [6], an approach is presented that aims to accelerate ghost removal by identifying regions with ghosting and reinitializing those regions. However, it still cannot completely eliminate ghost regions.

This paper presents several improvements based on the ViBe algorithm. These enhancements involve refining the background modeling approach and incorporating WLBP features to suppress ghost regions. Additionally, an adaptive factor is introduced to dynamically update the radius threshold according to the degree of scene variation, thereby adapting to environmental changes. Finally, post-processing using morphological operations is applied to meet the requirements for accurate motion object detection.

15.2 Method Flow

The improved ViBe algorithm workflow is illustrated in Fig. 15.1. Initially, a more sophisticated modeling method, the five-frame differencing technique, is employed to acquire the 'true background.' The WLBP algorithm is then used to extract texture features from the frame difference images for background model initialization. In the face of scenes with varying complexities, an adaptive radius threshold is set based on the Euclidean distance between current frame pixels and their background sampling data. This threshold helps determine differences between images. If an image is classified as foreground, it is binarized; otherwise, the background model is updated. Finally, the resulting binary image undergoes a morphological operation known as closing, involving dilation followed by erosion, to obtain a complete foreground image.

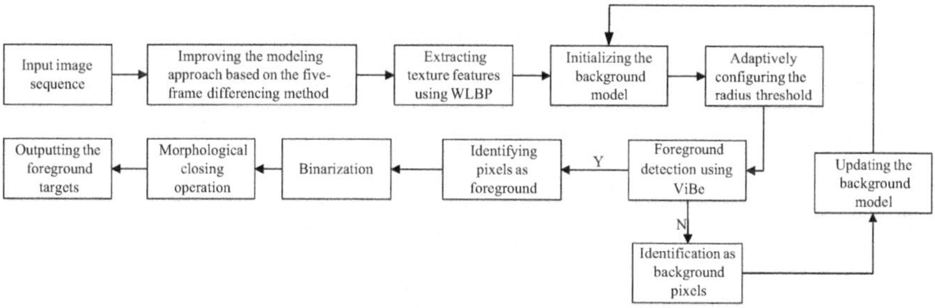

Fig. 15.1 Improved ViBe algorithm workflow

15.3 Improved ViBe Algorithm

15.3.1 Improved Modeling Approach

The ViBe algorithm establishes a background model based on the first frame of the video. For each pixel in the current frame, a set of samples, denoted as $M\ (k)$ is created by randomly selecting k pixels from the background model. The data in the $M\ (k)$ is randomly generated within the eight-neighborhood data.

$$M(k) = \{\ v_1, v_2, v_3 \cdots v_i \cdots v_N\} \tag{15.1}$$

where v_i represents the randomly sampled values from the 8-neighborhood of a random pixel, $i = 1, 2, \ldots, N$.

In the case where the first frame of a video sequence may contain moving objects, the ViBe algorithm encounters a challenge when constructing the background model using this initial frame. This issue arises when originally stationary objects begin to move, causing the ViBe algorithm to mistakenly classify regions covered by the moving objects as dynamic areas. Consequently, ghost regions can emerge during subsequent detection processes. To mitigate the presence of ghost regions and establish a more accurate background model, this algorithm utilizes a continuous five-frame differencing method [7]. Notably, the frame differencing technique possesses the advantage of being less affected by lighting conditions, which accelerates the elimination of ghost regions. The process begins by extracting five consecutive frames and denoting them as: $P_{t-2}(x, y)$, $P_{t-1}(x, y)$, $P(x, y)$, $P_{t+1}(x, y)$, $P_{t+2}(x, y)$. Additionally, an automated threshold selection, denoted as T is applied. Then, the algorithm computes differences between the middle frame and the preceding two frames, as well as the following two frames, as follows:

$$F_{t,t-2} = \begin{cases} 1 & \left|P(x, y) - P_{t-2}(x, y)\right| > T \\ 0 & \left|P(x, y) - P_{t-2}(x, y)\right| \le T \end{cases} \tag{15.2}$$

$$F_{t,t-1} = \begin{cases} 1 & \left|P(x, y) - P_{t-1}(x, y)\right| > T \\ 0 & \left|P(x, y) - P_{t-1}(x, y)\right| \le T \end{cases} \tag{15.3}$$

$$F_{t,t+1} = \begin{cases} 1 & \left|P(x, y) - P_{t+1}(x, y)\right| > T \\ 0 & \left|P(x, y) - P_{t+1}(x, y)\right| \le T \end{cases} \tag{15.4}$$

$$F_{t,t+2} = \begin{cases} 1 & \left|P(x, y) - P_{t+2}(x, y)\right| > T \\ 0 & \left|P(x, y) - P_{t+2}(x, y)\right| \le T \end{cases} \tag{15.5}$$

Performing logical 'AND' operations between $F_{t,t-2}$ and $F_{t,t-1}$, as well as between $F_{t,t+1}$ and $F_{t,t+2}$, yields the corresponding images $F_{51}(x,y)$ and $F_{52}(x,y)$.

$$F_{51}(x, y) = F_{t,t-1}(x, y) \cap F_{t,t-2}(x, y) = \begin{cases} 1 & F_{t,t-1}(x, y) = 1 \; and \; F_{t,t-2}(x, y) = 1 \\ & 0 \; others \end{cases} \tag{15.6}$$

$$F_{52}(x, y) = F_{t,t+1}(x, y) \cap F_{t,t+2}(x, y) = \begin{cases} 1 & F_{t,t+1}(x, y) = 1 \; and \; F_{t,t+2}(x, y) = 1 \\ & 0 \; others \end{cases} \tag{15.7}$$

The fusion of $F_{51}(x,y)$ and $F_{52}(x,y)$ results in the final motion object image $F(x,y)$, establishing an authentic background.

15.3.2 Extract Texture Features from the Frame Difference Images by WLBP

Since the ViBe algorithm primarily relies on pixel differences for object detection, it may struggle with accurately detecting objects with weak texture information and exhibit sensitivity to changes in lighting conditions. In cases where lighting variations are present in the scene, this can lead to inaccurate background estimation or detection errors. To enhance the accuracy of motion object extraction, mitigate the presence of ghost regions, and overcome susceptibility to lighting conditions, an improved LBP operator—weighted local binary patterns (WLBP)—is proposed during the background modeling phase [8]. The WLBP algorithm is applied to the difference images obtained after five-frame differencing, describing each frame using WLBP values. The concept of weighting is introduced, with weights calculated based on pixel difference values. This approach allows for a more precise capture of texture information in the images, facilitating better discrimination between foreground and background.

First, select the image region from which features are to be extracted. Choose a central pixel and define the LBP radius $r = 1$ and the number of neighboring pixels $P = 8$. For the surrounding pixels of the central pixel, calculate their grayscale differences with respect to the central pixel, and compute weights based on these difference values.

$$W = e^{-d^2/2\sigma^2} \tag{15.8}$$

Here, d represents the distance between neighboring pixels and the central pixel, and σ denotes the standard deviation of the Gaussian function. Based on the calculated weights and grayscale difference values, the surrounding pixels are partitioned into different regions. A threshold is selected for each region, and the LBP encoding is weighted using these thresholds to obtain the WLBP feature.

$$WLBP(x_k, y_k) = \sum\nolimits_{i=0}^{i-1} W\left(I_{(x_i,y_i)} - I_{(x,y)}\right) \times 2^i = \begin{cases} 1, \left|I_{(x_i,y_i)} - I_{(x,y)}\right| \geq T_h \\ 0, \left|I_{(x_i,y_i)} - I_{(x,y)}\right| < T_h \end{cases} \tag{15.9}$$

where W is the weight, $I_{(x,y)}$ is the grayscale value of the central pixel, $I_{(x_i,y_i)}$ is the grayscale value of the neighboring pixel, and i is the position of the binary bit. For each region, it is encoded as a binary number based on the comparison result of the grayscale difference between the surrounding pixels and the central pixel with the threshold. Usually, 1 is used to represent greater than or equal to the threshold, and 0 is used to represent less than the threshold.

By combining frame differencing and the WLBP algorithm, it is possible to achieve more precise background modeling. Frame differencing provides pixel-level foreground information, while the WLBP algorithm describes texture features. The introduction of weights into the ViBe algorithm, combined with these two techniques, enhances the accuracy and robustness of background modeling, effectively suppressing the presence of ghost regions.

15.3.3 Adaptive Radius Threshold Foreground Segmentation

For each pixel in a new input frame, a comparison is made with its background model. If the difference between this pixel and its background model exceeds a certain threshold over consecutive frames, and this threshold is surpassed a predetermined number of times, the pixel is classified as a foreground pixel [9].

As shown in Fig. 15.2, C_1 and C_2 are part of the two-dimensional color space (C_1, C_2). The pixel value of pixel k in the current frame is $v(k)$. $v(k) \bullet S_{(R)}(v(k))$ is a region with $v(k)$ as the center and R as the radius. The number of intersection points between the sample set $M(k)$ and $S_{(R)}(v(k))$ is denoted as ∂, which represents the similarity between $v(k)$ and the sample set $M(k)$. ∂ is calculated as follows:

$$\partial = \left\{S_{(R)}(v(k)) \cap M(k)\right\} \tag{15.10}$$

When ∂ is greater than a threshold value, v(k) is classified as background, and when ∂ is less than the threshold, v(k) is classified as foreground. The binary result is represented as:

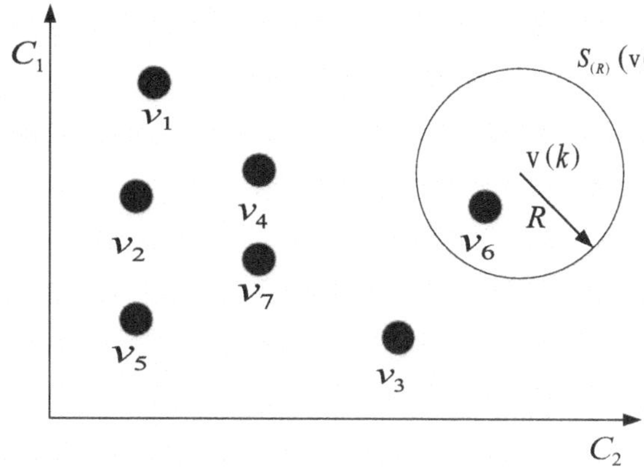

Fig. 15.2 Illustration of two-dimensional Euclidean color space classification

$$F(x, y) = \begin{cases} 1, & \partial < th \\ 0, & others \end{cases} \tag{15.11}$$

The thresholds employed in the ViBe algorithm are typically empirical values, where a fixed threshold is used for binarizing an image [10]. This approach is primarily suitable for static scenes, while real-world scenarios are seldom entirely static. To address specific situations, such as high-dynamic areas (e.g., swaying tree branches or rippling water surfaces), thresholds can be increased to avoid detecting foreground objects that cannot be directly measured. Conversely, for scenarios with lower dynamics, thresholds can be reduced to monitor subtle changes in the foreground. In order to adapt to the complexity of test scenarios and prevent false positives or false negatives, this paper optimizes the background modeling and threshold selection process. This optimization enables adaptive adjustments R based on the scene's complexity.

First, define an adaptive factor η to characterize the complexity of the scene.

$$\eta = \frac{\mu}{N}\sum_{i=1}^{n} |I_x - P_x| \tag{15.12}$$

where Ix represents the pixel value of the current frame pixel x, Px is the background sampling factor for pixel x, and μ is the scaling factor of η. A higher numerical value implies a greater scene variation, and consequently, the threshold for such scenarios should also be higher to reduce the influence of these factors. Conversely, the threshold should be lower for lower values. Secondly, the radius threshold R dynamically adjusts the update rate of the background model θ based on the adaptive factor η.

$$R(x, y) = \begin{cases} R(x, y)(1 - \theta), R(x, y) > \eta \\ R(x, y)(1 + \theta), R(x, y) > \eta \end{cases} \tag{15.13}$$

Here, θ represents the update rate of the background model and $R(x, y)$ denotes the radius threshold at pixel location (x, y).

Finally, following the initialization at R, foreground segmentation is carried out using the conventional ViBe algorithm's foreground segmentation method.

$$F(x, y) = \begin{cases} 1, \sharp\{dis(I_x, P_x) < R\} < th \\ 0, others \end{cases} \tag{15.14}$$

Here, $F(x, y)$ represents the foreground binary image, $dis(Ix, Px)$ is the Euclidean distance, and th is the minimum match count.

15.3.4 Foreground Detection

Due to the presence of holes in the obtained foreground binary image $F(x, y)$, this paper employs morphological binary closing operations [11] to effectively remove small noise and fill the holes in the image. This approach achieves the desired result of relatively complete foreground object detection.

$$F(x, y) \cdot B = (F(x, y) \oplus B) \ominus B \tag{15.15}$$

In the equation, the square structural element B defines morphological operations, where • represents binary closing, $\oplus$ denotes dilation, and $\ominus$ signifies erosion.

15.4 Experimental Results and Analysis

To validate the performance of our proposed method in eliminating ghost regions, adapting to varying lighting and dynamic scenes, and assessing overall algorithmic capabilities, we conducted simulation experiments in the MATLAB R2018b environment. The experimental platform consisted of an i7-9750H 2.6 GHz CPU with 16 GB of RAM, running on Windows 10, with Visual Studio 2019 and OpenCV 4.2.0. Video data were selected from partial videos in the CDW2014 dataset and videos captured by the authors. The algorithm parameters were set as follows: background modeling sampling elements $N = 20$, threshold $T_h = 20$, minimum match count $R_{min} = 2$, adaptive factor η, scaling factor $\mu = 2$, and update rate $\theta = 1$. In the same environment, we compared our method with GMM, KDE algorithms, and the traditional ViBe algorithm.

Fig. 15.3 Pedestrian video detection results. (**a**) The original frame image; (**b**)ViBe algorithm; (**c**) proposed algorithm

15.4.1 Ghost Region Suppression Experiment

Figure 15.3 presents the test results for the 27th, 103rd, and 205th frames of the video sequence. As the number of video frames increases, the background model is continuously updated to adapt to scene variations. For the ViBe algorithm, the results exhibit significant ghost regions in the 27th frame; however, over time, they gradually diminish in the 103rd and 205th frames, although some residual ghost regions remain. By comparing Fig. 15.3b, c, it is evident that our proposed method effectively suppresses ghost regions, allowing for better discrimination between foreground and background, reducing false positive detections, and achieving accurate extraction of moving objects.

15.4.2 Experiments Under Illumination Variations and Dynamic Scenes

Figure 15.4 presents the test results for the 'Fall' video in a dynamic scene with lighting variations and tree leaf motion interference. The ViBe algorithm mistakenly detects moving background elements like swaying leaves as foreground, leading to inaccurate extraction of moving objects. In comparison to our method, it is evident that the ViBe algorithm exhibits significant false positives, while our approach effectively distinguishes between moving objects (such as cars and pedestrians) and swaying tree leaves, resulting in accurate target extraction with a lower false positive

(a) 211th frame

(b) 292th frame

(c) 513th frame

Fig. 15.4 Dynamic scene video test results. (**a**) The original frame image; (**b**) ViBe algorithm; (**c**) proposed algorithm

rate. Our method demonstrates superior performance in handling lighting variations and exhibits clear advantages in terms of reliability in complex and dynamically changing environments, as well as effectiveness in object detection.

15.4.3 Algorithm Comprehensive Performance Experiment

To evaluate the overall performance of our proposed method, Fig. 15.5 presents the results of comparisons with the KDE algorithm, GMM, and ViBe algorithm across multiple video tests.

Through comparative analysis, it can be observed that the KDE algorithm tends to produce extensive false negatives in motion target detection, making it prone to missing actual moving objects. The GMM algorithm is susceptible to lighting changes during the detection process, resulting in erroneous detection results when lighting conditions change, leading to false positives or false negatives. The ViBe algorithm performs poorly in dynamic scenes, especially when there are numerous dynamic elements in the background or significant scene changes occur. In such cases, the ViBe algorithm may incorrectly identify these dynamic elements as targets, resulting in large false positive regions. Both the GMM and ViBe algorithms, while capable of detecting complete regions of moving targets, are still affected by noise and exhibit slight target omissions. In contrast, our proposed method consistently and accurately detects targets under various conditions, exhibiting minimal false positives and false negatives. It remains unaffected by lighting

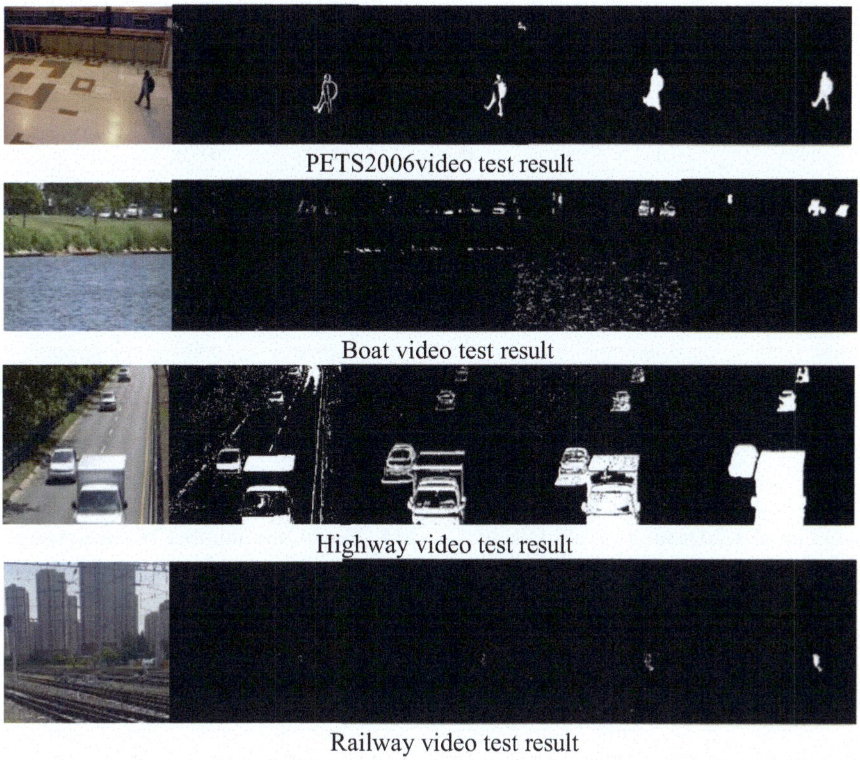

Fig. 15.5 The test results of four algorithms in different complex scenarios. (**a**) Input image; (**b**) KDE algorithm; (**c**) GMM; (**d**) ViBe algorithm; (**e**) proposed algorithm

changes, dynamic scenes, or noise, demonstrating superior performance in motion target detection, as evident from the comparative results.

To effectively assess the 'quality' of detections made by the various algorithms, a quantitative study on the stability of our proposed method in comparison to the GMM, KDE algorithm, and ViBe algorithm was conducted using three metrics: accuracy ($P_{\rm re}$), recall rate ($R_{\rm e}$), and $F - M$ [12]. The specific definitions of these metrics are as follows:

$$P_{re} = \frac{P_{TP}}{P_{TP} + P_{FP}} \tag{15.16}$$

$$R_e = \frac{P_{TP}}{P_{TP} + P_{FN}} \tag{15.17}$$

$$F - M = \frac{2 \times P_{re} \times R_e}{P_{re} + R_e} \tag{15.18}$$

Table 15.1 Comparison of P_{re} indicators

Test videos	KDE algorithm	GMM	ViBe	Proposed method
PETS2006	0.716	0.877	0.811	0.927
Boat	0.624	0.693	0.649	0.849
Highway	0.577	0.750	0.792	0.905
Railway	0.795	0.774	0.808	0.916

Table 15.2 Comparison of R_e indicators

Test videos	KDE algorithm	GMM	ViBe	Proposed method
PETS2006	0.743	0.906	0.767	0.955
Boat	0.580	0.783	0.690	0.913
Highway	0.695	0.857	0.754	0.872
Railway	0.818	0.829	0.841	0.964

where P_{TP} represents the number of foreground pixels correctly detected, P_{FP} is the number of foreground pixels falsely detected, P_{TN} is the number of background pixels correctly detected, and P_{FN} represents the number of background pixels falsely detected. P_{re} is the ratio of correctly classified foreground pixels to the total foreground pixels, R_e is the ratio of correctly detected foreground pixels, and F–M is the harmonic mean of the first two metrics. When the numerical results are closer to 1, they indicate higher precision and recall rates.

Tables 15.1, 15.2, and 15.3 provide the detection performance metrics for the four algorithms. The KDE algorithm fails to correctly detect targets in some cases, resulting in large areas of voids that lead to lower accuracy and recall rates and significant instances of missed detection. Both GMM and the traditional ViBe algorithm exhibit relatively good accuracy in their detection results, but due to their sensitivity to lighting and dynamic scenes during the detection process, they experience false positives and false negatives, resulting in higher false positive rates and false negative rates. In various test scenarios such as subway stations, boats, highways, and railways, the KDE algorithm exhibits an accuracy difference of 21.1, 22.5, 32.8, and 12.1% compared to our method, while the recall rates differ by 21.2, 33.3, 17.7, and 14.6%, respectively. The comprehensive evaluation metric F-M value shows differences of 0.193, 0.264, 0.263, and 0.136. In comparison, the GMM algorithm has accuracy differences of 5, 15.6, 15.5, and 14.2% compared to our method, with recall rate differences of 4.9, 13, 15, and 13.5%, respectively. The comprehensive evaluation metric F–M value exhibits differences of 0.031, 0.13, 0.083, and 0.117. Compared to the other three detection algorithms, our method demonstrates more accurate detection of moving targets, with a significant advantage in accuracy.

Table 15.4 presents a comparison of the average processing time per frame for the four algorithms in different scenarios. The KDE algorithm requires more computational resources and time as it needs to maintain and update the kernel density estimates for each pixel. In the four different test scenarios, both KDE

Table 15.3 Comparison of *F-measure* indicators

Test videos	KDE algorithm	GMM	ViBe	Proposed method
PETS2006	0.729	0.891	0.788	0.922
Boat	0.601	0.735	0.669	0.865
Highway	0.630	0.800	0.772	0.883
Railway	0.805	0.800	0.824	0.941

Table 15.4 The average processing time (ms/frame) of the four algorithms for each frame of image is different

Test videos	KDE algorithm	GMM	ViBe	Proposed method
PETS2006	117.213	66.825	39.021	47.873
Boat	92.047	41.082	21.380	25.419
Highway	105.620	49.367	23.674	28.165
Railway	80.219	37.048	17.932	22.517

and GMM algorithms have an average processing time that is 68 and 18 ms higher than our method, respectively. Our method's average processing time per frame is slightly higher than ViBe algorithm but significantly lower than KDE and GMM. This indicates that our method offers a balance between algorithm complexity and real-time detection requirements for moving targets. It suggests that this method can achieve good results in practical application scenarios.

15.5 Conclusion

We proposed a motion target detection method that combines the five-frame difference method with the weighted local binary patterns (WLBP) algorithm, and incorporates an adaptive radius threshold improvement into the ViBe algorithm. The five-frame difference method determines potential motion regions by comparing the differences between consecutive frames of five images. The WLBP algorithm is a texture feature extraction method commonly used for recognizing texture information in objects. By integrating these two methods, we can analyze images more comprehensively and construct a background model for precise motion target extraction. The introduction of the adaptive radius threshold allows the detection parameters to be adjusted based on specific conditions and lighting situations in the scene, rather than relying on fixed thresholds. This adaptive approach enables the algorithm to adapt to various changes in complex environments, thereby enhancing detection accuracy. The results show that the improved ViBe algorithm has significant advantages in suppressing ghost regions, reducing false positives and false negatives caused by lighting and dynamic environments, and improving the erosion phenomenon of foreground targets. It achieves more accurate motion target detection in complex environments. Furthermore, this method demonstrates improved adaptability compared to the traditional ViBe algorithm, delivering excellent detection performance and meeting the demands of motion target detection in complex environments. It holds high practical value and prospects for application in the field of communication image processing and analysis.

Foundation Item Gansu Province University Industry Support Plan Project (2023CYZC-40), National Natural Science Foundation Project (61761027), Gansu Province Excellent Postgraduate 'Innovation Star' Fund Project (2023CXZX-546).

References

1. Li, S.J., Li, Z.M., Li, W.: An overview of improved ViBe algorithms. In: 2021 7th International Conference on Computer and Communications (ICCC), p. 9674734. IEEE, Chengdu (2021)
2. Biao, W., Lin, Z.: Improvements on Vibe algorithm for detecting foreground Objects. In: Proceedings of 2016 5th International Conference on Computer Science and Network Technology (ICCSNT), pp. 40–44 (2016)
3. Q. Liu, Z. Li , S. Li. Performance analysis on Vibe detection of moving object. 2022 International Conference on Automation, Robotics and Computer Engineering, ICARCE, Wuhan, 1–5 (2022)
4. Liu, J., Zhang, Y., Zhao, Q.: Adaptive ViBe algorithm based on Pearson correlation coefficient. In: 2019 Chinese Automation Congress, CAC, pp. 4885–4889 (2019)
5. Shao, X., Chen, X., Li, K., et al.: An improved moving target detection method based on Vibe algorithm. In: 2018 Chinese Automation Congress, CAC, pp. 1928–1931, Xi'an (2018)
6. Bai, E., Zhao, J., Wu, Y.: An improved ViBe algorithm for quickly suppressing ghosts. In: 2020 IEEE International Conference on Advances in Electrical Engineering and ComputerApplications(AEECA), pp. 621–624, Dalian (2020)
7. Qiming, Z., ChengQian, M.: A vehicle detection method in tunnel video based on ViBe algorithm. In: 2017 IEEE 2nd Advanced Information Technology, Electronic and Automation Control Conference, IAEAC, pp. 1545–1548, Chongqing (2017)
8. Zhang, J., Xiao, X.: Face recognition algorithm based on multi-layer weighted LBP. In: 2015 8th International Symposium on Computational Intelligence and Design (ISCID), pp. 196–199, Hangzhou (2015)
9. Peng, D.G., Liu, W.W., Qi, E.J., et al.: Liquid leakage detcetion of power plant pipelines based on improved ViBe algorithm. In: 2021 Power System and Green Energy Conference (PSGEC), p. 21136784. IEEE, Shanghai (2021)
10. Zhao, D., Tan, J., Yang, W., et al.: An improved VIBE algorithm for fast suppression of ghosts and static objects. In: 2018 IEEE International Conference on Mechatronics and Automation, ICMA, pp. 889–893, Changchun (2018)
11. Gao, J., Zhu, H.: Moving object detection for video surveillance based on improved ViBe. In: 2016 Chinese Control and Decision Conference, CCDC, pp. 6259–6263, Yinchuan (2016)
12. Adigüzel, D., Özyilmaz, L.: Improving of ViBE algoritihm for moving target detection at maritime surveilance. In: 2018 26th Signal Processing and Communications Applications Conference, SIU, pp. 1–4, Izmir (2018)

Chapter 16
Constructing a Neural Network Model for Traditional Chinese Medicine's "Shanghan Lun"

Ching-Wen Chen and Long-Chen Hung

Abstract This study coupled both new systems of syndrome essential factor and formula essential factor with the artificial intelligence technique to analyze all the formulas of the Shang Han Lun (伤寒论) to build an AI model of formula-syndrome essential-factor correspondence. It is developed by self-coding with the software of Microsoft Visual Studio C#.NET to design all the functions that we need. One of the methods called back propagation network (BPN) of the artificial neural network technique was employed to build the AI model. This study employed the book of Shang Han Lun (伤寒论) for the materials to undertake the structural analyses. The mathematical model can then be applied to the practical prediction of prescription. That demonstration shows that it is feasible and practical to forecast the precise formulas for the patients. However, some divergence appears when the amount of neuron nodes is too large. Two methods were employed to improve the efficiency of convergence, i.e., the adjustment of learning rate and the conjugate gradient method.

16.1 Introduction

The standard of "National Standard of the People's Republic of China – Classification and Code of Diseases and Syndromes of Traditional Chinese Medicine – Name and Classification Code of Syndromes of Traditional Chinese Medicine" contains six categories and 258 subcategories, among which the etiological syndrome category (病因证候类) has 102 subcategories, the yin-yang-qi-blood-fluid-phlegm syndrome category (阴阳气血津液痰证候类) has 63 subcategories, the zang-fu-meridian syndrome category (脏腑经络证候类) has 62 subcategories, the

C.-W. Chen (✉)
Department of Business Management, Taipei University of Technology, Taipei, Taiwan
e-mail: ccw@ntut.edu.tw

L.-C. Hung
Department of Electrical Engineering, Cheng Shiu University, Kaohsiung, Taiwan
e-mail: 0677@gcloud.csu.edu.tw

S.-P. Tseng et al. (eds.), *Advances in Intelligent Information Hiding and Multimedia Signal Processing, Volume 2*, Smart Innovation, Systems and Technologies 416, https://doi.org/10.1007/978-981-97-8760-9_16

six-meridian syndrome category (六经证候类) has 19 subcategories, the wei-qi-nutritive-blood syndrome category (卫气营血证候类) has 9 subcategories, and other syndromes category, etc. However, there are significant differences between various standards of traditional Chinese medicine syndromes, which cause confusion for practitioners, because the standards of different standards are not consistent.

In recent years, the research on traditional Chinese medicine syndromes has made progress, especially the proposal of "syndrome elements". This new concept has a profound impact on the application of traditional Chinese medicine diagnosis and treatment, principles and methods, prescriptions and drugs, etc. The classification units of syndrome elements include disease location elements and disease nature elements. For complex syndrome systems, using syndrome elements for classification is simpler and easier to master, and fewer syndrome elements can also provide a lot of information about diseases.

Professor Wang Yongyan published "Suggestions on Improving the System of Diagnosis Methods in Traditional Chinese Medicine" in 2003 [1–3]. Later, Zhang Zhibin and Wang Yongyan further proposed in 2007 that the system of traditional Chinese medicine syndromes is a complex nonlinear multidimensional multilevel system [4]. They emphasized that it is necessary to introduce the concept of complex science and establish a new system that meets the modern standard requirements and can adapt to the complex and dynamic clinical situations of syndromes.

In addition to the significant progress made in research centered around syndrome elements mentioned above, the proposal of "formula elements" has further enriched the integrity of the new system of TCM syndrome differentiation and treatment. Li Yuhang proposed the concept of formula elements in 2009 [5], and based on the principle of "formula-syndrome correspondence" for syndrome differentiation and treatment, put forward the principle of "formula-syndrome element correspondence" for formula composition. Therefore, "formula-syndrome element correspondence" is an important reference principle during the process of clinical syndrome differentiation and formula composition. Determining the corresponding formula elements based on syndrome elements is one of the principles for TCM formula composition, which is also known as the "formula-syndrome element correspondence" principle. This principle has strong practicality and operability in clinical application, and if it is organically combined with the principle of "jun-chen-zu-shi 君臣佐使", it will have positive significance in enriching the theory of formula compatibility in TCM pharmacology.

The current clinical medication principles for syndrome differentiation and treatment can be mainly summarized into four categories:

1. The principle of "formula-syndrome correspondence 方证相应": This is one of the shortcuts for accurately and effectively using ready-made prescriptions and is also key to the clinical effectiveness of formulas. There is no difference between ancient and modern prescriptions in terms of quality as long as they correspond to the syndrome.

2. The principle of "optimization through jun-chen-zu-shi 君臣佐使": The entire formula controls the status and function of each medicinal ingredient through the relationship between "jun", "chen", "zu", and "shi", indicating that this method is highly planned.
3. The principle of "correspondence between drug and syndrome 药证相应": Drug indications are evidence for clinical medication in TCM. Research on drug indications aims to establish a normative standard for clinical medication in TCM.
4. The principle of "formula-syndrome element correspondence 方-证要素对应" : This is a recently proposed principle by Professor Li Yao [5], which is based on a correspondence relationship between "formula elements" and "syndrome elements".

There are several categories of data mining methods used for research on TCM syndromes, including:

1. Association rules: Wang et al. [6], Zhang et al. [7]
2. Cluster analysis: Yuan et al. [8]
3. Artificial Neural Network: Li et al. [9, 10]
4. Decision tree: Liu and Ni [11]
5. Bayesian network: Wang et al. [12]

Based on the significant research progress made in recent years on the new system of TCM syndrome differentiation and treatment mentioned above, this study aims to establish a system for syndrome differentiation and treatment of Shanghan Lun (伤寒论) prescriptions, namely, "syndrome elements, formula elements system, formula-syndrome element correspondence".

This study also proposes a mathematical model to simulate the formula-syndrome element correspondence relationship in Shanghan Lun (伤寒论) prescriptions. However, the formula-syndrome element correspondence equation Yj = f(xi) in Shanghan Lun (伤寒论) is itself a multivariate nonlinear vector equation. We have chosen artificial neural network technology as the mathematical model to simulate the "formula-syndrome element correspondence" in Shanghan Lun (伤寒论) because it is currently an excellent solution for dealing with multivariate nonlinear systems.

The purpose of this study is to combine the latest research progress on "syndrome elements and formula elements" with artificial intelligence technology in the computer field to study the prescriptions of Shanghan Lun (伤寒论) and establish a mathematical model that can handle the "formula-syndrome element correspondence and main syndrome-drug correspondence" in Shanghan Lun (伤寒论).

The objectives to be completed are as follows:

1. Analyze the structure of Shanghan Lun (伤寒论) prescriptions and establish a system of "syndrome elements, formula elements, formula-syndrome element correspondence, and main syndrome-drug correspondence".

2. Use artificial neural networks to establish the mathematical model of the "main syndrome-drug correspondence" of Shanghan Lun (伤寒论).
3. Input a large amount of data on "main syndrome-drug correspondence" into the neural network model, allowing the system to constantly learn, so that it can obtain the weights and thresholds of each coefficient. The mathematical model of "main syndrome-drug correspondence" in Shanghan Lun (伤寒论) can then have the function of calculating and predicting prescriptions. It has certain reference value for assisting clinical diagnosis and treatment and can provide methodological reference for better inheriting.

16.2 Research Methods and Procedures

16.2.1 The Tools Used

1. Algorithm program for self-developed neural network (ANN) and its man-machine interface platform.
2. Computer technology required for program development process, including using Microsoft Visual Studio C#.NET and Access database to write programs to develop the system.
3. Using one of the mathematical algorithms of neural network (ANN): Back propagation network (BPN) to establish the mathematical model of "formula-syndrome element correspondence" system in Shanghan Lun (伤寒论).

16.2.2 Analysis of Shanghan Lun (伤寒论)

This study first carries out a structured analysis of prescriptions in Shanghan Lun (伤寒论) based on the "Eleventh Five-Year Plan" textbook of Shanghan Lun (伤寒论) Lecture Notes, and establishes the system of syndrome elements, formula elements, formula-syndrome element correspondence, and main syndrome-drug correspondence. Various corresponding information can be obtained from the large-scale data of structured analysis of prescriptions in Shanghan Lun (伤寒论).

16.2.3 Classification Basis of Drug Properties

The classification basis of drug properties is based on the content of Chinese Pharmacopoeia 2005 when analyzing the system of formula elements and formula-syndrome element correspondence.

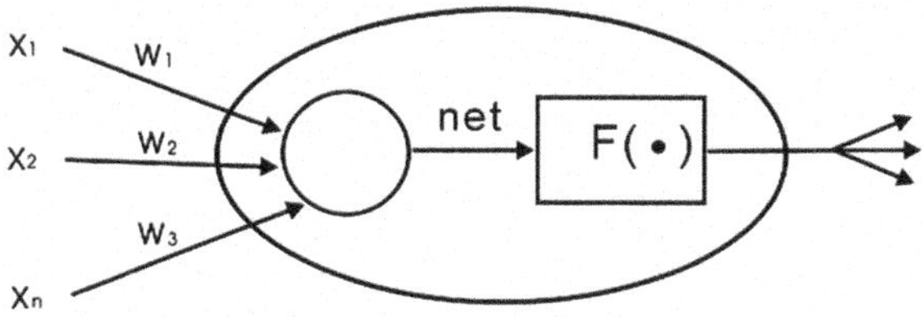

Fig. 16.1 The whole neuron model structure

16.2.4 Using BP Back Propagation Network (BPN)

The concept of artificial neural network (ANN) is derived from biological neural networks, and its logical composition and structure are similar to those of biological neural networks. ANN also has the characteristics of human brain function, including learning, recalling, and reasoning. The inspiration of ANN comes from the study of brain neurons. The relationship between the output value and input value of a neuron can be expressed as a formula, which can be derived as Fig. 16.1.

$$Y_j = f\left(\sum w_{ij}x_i - \theta_j\right) \tag{16.1}$$

Y_j: represents the output signal mimicking the jth biological neuron (i.e., the output variable, a key element in the formulation),
F: denotes the transformation function imitating the biological neuron model,
w_{ij}: symbolizes the synaptic strength between the ith and jth biological neurons (i.e., the connection weight),
x_i: signifies the input signal from the ith biological neuron (i.e., the input variable, an essential element for diagnosis),
θ_j: is the threshold of the jth biological neuron (i.e., the threshold value),

In other words, once the weights Wij and thresholds θj of the entire system of formula-syndrome elements in Shanghan Lun (伤寒论) can be calculated through learning from a large number of cases, they can be applied to clinical practice to estimate the optimal drug prescription.

Our model is based on a back propagation network (BPN), and the core of this model is the multivariate nonlinear vector equation $Yj = f(x_i)$, which defines the transformation from the input vector x_i to the output Y_j. This transformation involves a combination of weighted sums and activation functions.

Each layer in the network performs a computation that involves calculating the weighted sum of its inputs, followed by the application of an activation function. The weights and biases in the model are initialized and then updated through the training process using backpropagation. The specific activation functions ReLU used in our model are chosen based on their performance and suitability for the problem at hand.

How can we determine that we have found the optimal solution or the weights W_{ij} and thresholds θ_j when writing a computer program to simulate the entire neural network system? This requires calculating the root mean square error function

(Error Function). If this error function tends to converge, then the optimal solution has been found. The Error Function is shown as follows:

$$E = \frac{1}{2}\sum \left(T_j - Y_j\right)^2 \tag{16.2}$$

T_j = the target output value of the output layer of the training example at the jth output unit.

Y_j = the inferred output value of the output layer of the training example at the jth output unit.

When the calculated root mean square error E is lower than the threshold value, it means that the standard solution has been obtained. Therefore, the learning process of artificial neural networks is to minimize the root mean square error function.

The operation process of the neural network used in this study can be divided into two stages:

1. Learning process: during the learning process, the network learns from examples according to a learning algorithm and adjusts the connection weights by repeated calculations;
2. Recalling process: during the recalling process, the network receives external input and recalls according to a recall algorithm. After repeated calculations, the output level neurons send out results. The learning process of feedforward neural networks is actually a supervised learning process, which obtains training examples and target output values from the problem domain and adjusts the connection weights by gradient steepest descent method. In computer program processing back propagation network, it obtains training examples and target output values from problem domain and inputs them into the network. Traditional back propagation uses gradient steepest descent method to adjust connection weights and deviation weights, but this often causes slow convergence speed. In this study, conjugate gradient method is used to accelerate convergence speed. The algorithm of back propagation network is divided into learning and recalling parts.

The reasons about choosing back propagation networks (BPN) over other types of neural networks are described as follows:

1. Problem Characteristics and BPN Adaptability: BPN is particularly suited for problems requiring supervised learning, where the network needs to learn from labeled data. This type of network is especially effective in establishing relationships between input and expected output, which is crucial in your research.
2. Gradient Descent and Weight Update Mechanism: BPN utilizes the gradient descent method to minimize prediction errors, which is very effective for approximating complex functions. Moreover, the weight update mechanism allows the network to learn and improve its predictions, necessary for your research goals.

3. Extensive Applications and Research Support: BPN has been widely studied and validated for its effectiveness in a variety of applications. This historical proof of its reliability and versatility is an important factor in choosing it for scientific research.
4. Comparison with Other Neural Networks: While networks like convolutional neural networks (CNNs) and recurrent neural networks (RNNs) may be more effective in specific applications, they can be overly complex or unsuitable in some situations. For example, CNNs are very powerful in processing image data, but if your data is not image-based, BPN might be a better choice.

Ease of Implementation and Flexibility: The architecture of BPN is relatively simple and easy to implement, and it can be adjusted according to specific needs, making it particularly useful for preliminary research or smaller datasets.

16.3 Results and Discussion

16.3.1 Structural Analysis of Prescriptions in Shanghan Lun (伤寒论)

This study conducted a structured analysis of the prescriptions in Shanghan Lun (伤寒论) based on the "Eleventh Five-Year Plan" textbook of Shanghan Lun (伤寒论) Lecture Notes [13], establishing the system of syndrome elements, formula elements, formula-syndrome element correspondence with a total of 109 different named prescriptions analyzed. Table 16.1 is the example.

Table 16.1 Structural analysis of prescriptions in Shanghan Lun (伤寒论)

Formulas name	Pathogenesis	Main syndrome	Drug composition	Syndrome elements	Therapeutic principles	Composition principles	Formula elements
1. 桂枝汤	风寒袭表	发热	桂枝	A 卫阳浮盛	A 解肌祛风	A 助卫阳散风邪	A 桂枝、生姜
	卫阳浮盛	汗出	生姜	B 营阴外泄	B 调和营卫	B 敛阴和营	B 芍药、大枣
	卫外不固	恶风	芍药			C 调补脾胃	C 甘草、生姜、大枣
	营阴外泄	头痛	大枣				
		脉浮缓	甘草				

16.3.2 Establishing Main Syndrome x Vector and Drug y Vector

The above table of main syndrome data and the table of drug composition data need to be converted into vector representations because the entire neural network performs calculations and obtains optimal solutions using vectors as the means.

16.3.3 Let the Neural Network Learn the Correlation Between These Main Syndrome x Vectors and Drug y vectors ectors to Calculate the Optimal Solution for Weights and Thresholds

The neuron model structure used in this study has a three-layer structure, which includes an input layer (main syndrome), a hidden layer, and an output layer (drugs). As there are a total of 354 main syndromes, the input layer x has 354 neural nodes. The number of hidden layers is set to 20 (in this study, it was found that setting it to 20 achieves the best convergence speed for the solution). Additionally, there are 98 drugs, so the output layer y has 98 neural nodes. There are a total of 118 training samples (Fig. 16.2).

16.3.4 Apply the Established Mathematical Model of Correspondence Between the Main Syndrome and Drug in Shang Han Lun to Actual Drug Prediction

As the number of inputted main syndrome x vectors and drug y vectors increases, and the length of the vectors becomes longer, the efficiency of the convergence process becomes increasingly poor, eventually leading to divergence and inability to converge. This is because as the sample size increases and the input vector dimensions become higher, the number of neural nodes in the input layer increases. In this study, the dimension of the input vector after complete input data is 354 dimensions (there are 354 main syndromes and 354 neural nodes), while the dimension of the output vector is 98 dimensions (there are 98 drugs and 98 neural nodes) (Fig. 16.3).

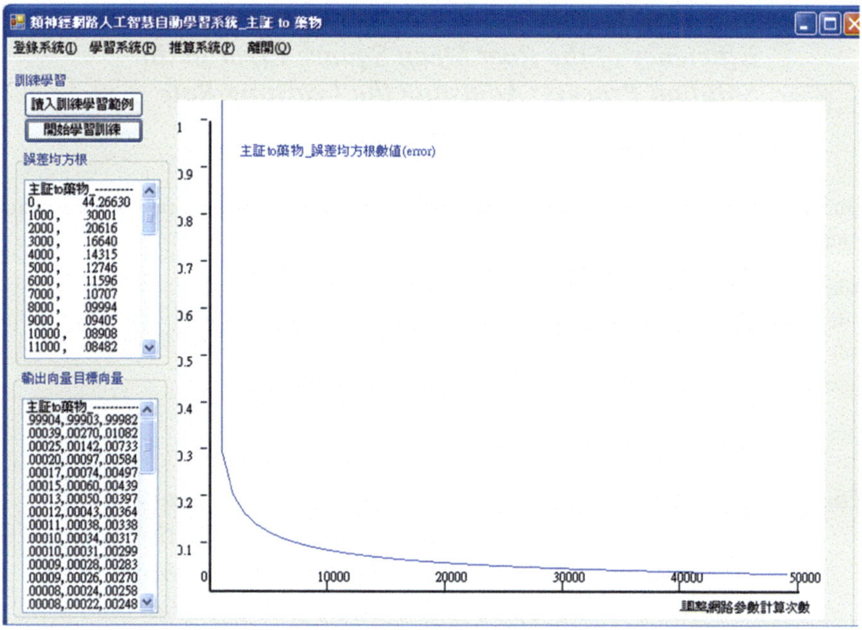

Fig. 16.2 The root mean square error function has converged, representing the optimal solution for weights and thresholds

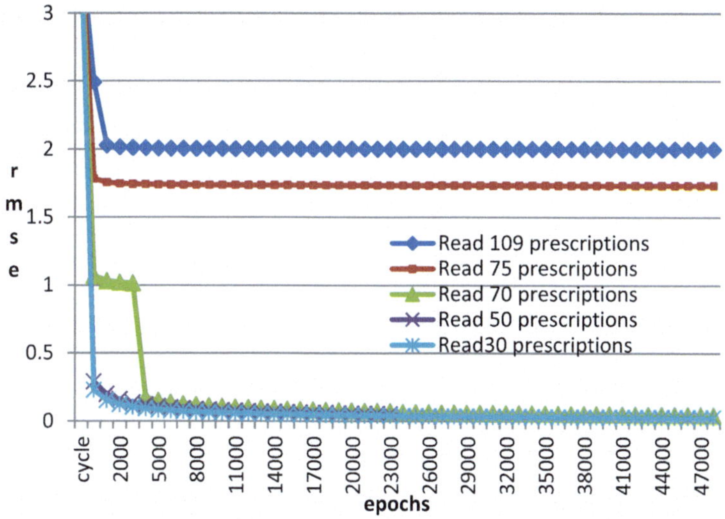

Fig. 16.3 Comparing the impact of various prescription reading amounts on the change in root mean square error of the main syndrome-drug system solution process

16.3.5 The First Method to Improve the Convergence Efficiency of the Root Mean Square Error in the Solution Process: Adjusting the Learning Rate Parameter

During the testing process of this study, the optimal learning rate parameters were found to be set to the following values:

rate_w = 0.95; // weight learning rate (input layer-hidden layer),
rate_w1 = 0.95; // weight learning rate (hidden layer-output layer),
rate_b1 = 0.95; // threshold learning rate for the hidden layer,
rate_b2 = 0.95; // threshold learning rate for the output layer.

16.3.6 The Second Method to Improve the Convergence Efficiency of the Root Mean Square Error in the Solution Process: Using Conjugate Gradient Method to Calculate the Optimal Solution of the System

Originally, we used the Steepest Descent Method for solving the equations in the program, but when the upper limit of the optimal solution was reached at reading 75 prescriptions, the system would fail to converge during the solution process. Therefore, to address this difficulty, we decided to use the conjugate gradient method.

After using conjugate gradient method to calculate the optimal solution of the system, we found that it did improve the convergence effect, but the system still encountered a failure to converge when too many Shang Han Lun prescriptions were read (Fig. 16.4).

16.4 Conclusion

This study conducts a structured analysis of Shang Han Lun prescriptions, focusing on creating mathematical models for the relationships between formula-syndrome elements and main syndrome-drug correspondences. The goal is to understand the qualitative and quantitative relationships between syndromes and drugs, aiding in clinical differentiation and identification. Results indicate that the method can accurately predict prescriptions and adjust for optimal weight and bias in the models. However, challenges arise with large neuron nodes, addressed by adjusting learning rates, applying conjugate gradient methods, and optimizing initial vectors. The study also deals with synonym issues in syndrome analysis, noting that the formula-syndrome essential-factor correspondence system is more efficient than

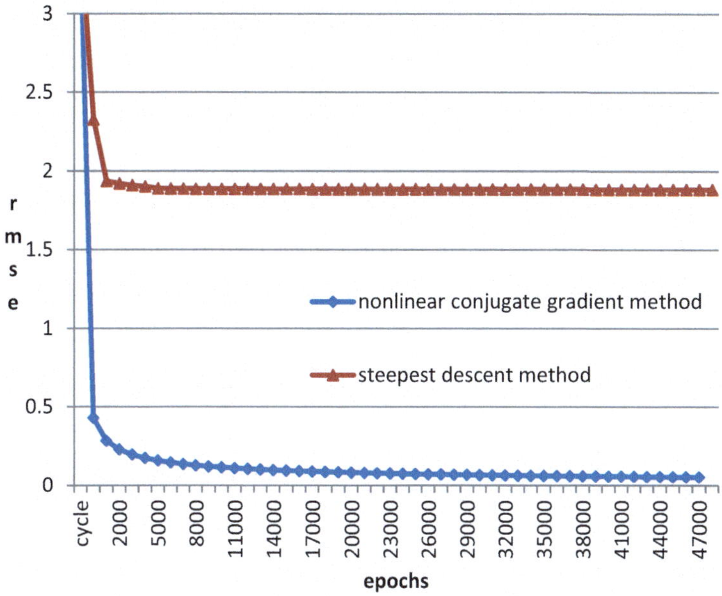

Fig. 16.4 Comparing the impact of using the steepest descent method and nonlinear conjugate gradient method on the change of root mean square error of the solution process when reading 88 prescriptions

the formula-syndrome correspondence system due to its simpler relationship and smaller neuron node dimension.

References

1. Wang, Y., Zhang, Z.: Review of syndrome nomenclature and classification research and presentation of hypotheses. J. Tradit. Chin. Med. **26**(2), 1–5 (2003)
2. Wang, Y.: Suggestions for improving the system of TCM syndrome differentiation methods. J. Tradit. Chin. Med. **45**(10), 729–731 (2004)
3. Wang, Y., Zhang, Z.: Revisiting several issues regarding improving the system of syndrome differentiation methods. J. Tradit. Chin. Med. **4**(1), 1–4 (2004)
4. Wang, Y., Zhang, Z.: Establishment of a new system for syndrome differentiation methods. J. Tradit. Chin. Med. **28**(1), 1–3 (2007)
5. Li, Y.: Discussion on "syndrome elements" and "formula elements". J. Tradit. Chin. Med. **24**(2), 117–121 (2009)
6. Wang, H., Hu, X.: Application of data mining based on association rules in clinical practice. J. Anhui Univ. (Nat. Sci. Edit.). **30**(2), 21–251 (2006)
7. Zhang, C., Yan, C., Song, L.: Mining algorithm for association rules in TCM kidney disease treatment information. J. Nat. Sci. Heilongjiang Univ. **22**(6), 842–845 (2005)
8. Lin, W.: Application of BP network in TCM differentiation of bi syndrome. J. Fujian College Tradit. Chin. Med. **7**(4), 41–43 (1997)

9. Li Jg, H.J., Yu, X.: Exploration of a quantitative diagnostic model for TCM syndromes based on neural networks. J. Henan College Tradit. Chin. Med. **20**(3), 6–8 (2005)
10. Li, J., Hu, J., Yu, X., Shen, J.: Exploration of a quantitative diagnostic model for TCM syndromes based on neural networks. J. Henan College Tradit. Chin. Med. **20**(118), 6–8 (2005)
11. Hu, J., Yan, S., Wang, X.: Artificial neural network classifier for pulse characteristics. J. China Med. Univ. **25**(6), 571–578 (1996)
12. Wang, X., Qu, H., Wang, J.: A quantitative diagnosis method for TCM based on data mining. J. Tradit. Chin. Med. **28**(1), 4–7 (2006)
13. Liu, D.: Shang han lun shi si jiang [Fourteen lectures on the Treatise on Cold Damage]. People's Medical Publishing House. (2013). ISBN 711717353X, 9787117173537

Chapter 17
Tumbleweed Algorithm for Multilevel Image Segmentation

Shu-Chuan Chu, Xiaomeng Yang, Jeng-Shyang Pan, Bin Yan, and Hongmei Yang

Abstract The tumbleweed algorithm (TA) is a new optimization algorithm inspired by a plant called tumbleweed, and like other evolutionary algorithms, this algorithm is used to solve parameter tuning in real-world problems. The algorithm simulates the process of tumbleweed from young shoots to maturity as well as the process of sowing seeds after maturity, and a growth cycle formulation is added to combine these two processes. In this paper, the TA algorithm is applied to the multilevel image segmentation problem, solving the problem on the basis of minimum cross-entropy thresholding. The experimental results show that the TA algorithm has good image segmentation results.

17.1 Introduction

In recent years, meta-heuristic algorithms have been heavily researched and have notable capabilities in a variety of application areas. Algorithm inventors often get algorithmic inspiration from plants and animals to develop metaheuristic algorithms with good performance, such as DE [1], TA [2], Particle Swarm Optimization (PSO) [3, 4], Artificial Bee Colony algorithm (ABC) [5], Cat Swarm Optimization [6], Butterfly Optimization Algorithm (BOA) [7], Cuckoo Search algorithm(CS) [8], Aquila Optimizer (AO) [9], Political Optimizer (PO) [10], Quasi Affine Transform Evolution (QUATRE) [11], Gannet Optimization Algorithm (GOA) [12], Pigeon Inspired Optimization (PIO) [13], and Whale Optimization Algorithm (WOA) [14]. These algorithms are usually grown into populations, which are constantly moved and computed in many iterations to find the optimal solution; meta-heuristic

S.-C. Chu · X. Yang · J.-S. Pan · B. Yan · H. Yang
College of Computer Science and Engineering, Shandong University of Science and Technology, Qingdao, China

J.-S. Pan (✉)
Department of Information Management, Chaoyang University of Technology, Taichung, Taiwan

S.-P. Tseng et al. (eds.), *Advances in Intelligent Information Hiding and Multimedia Signal Processing, Volume 2*, Smart Innovation, Systems and Technologies 416, https://doi.org/10.1007/978-981-97-8760-9_17

algorithms are often tested at CEC 2013 or CEC 2017 as a way to test the algorithmic capabilities.

Each meta-heuristic algorithm has its own unique evolutionary mechanism, but there are two phases of exploitation and exploration when performing optimization [15]. Exploration is to discover as many regions in the search space as possible that have an optimal solution [16], and the development phase is to locally search and converge in regions that may have an optimal solution [17]. Because the algorithm is stochastic, it is an important issue to be able to converge quickly to find the optimal solution [18].

Image segmentation is an important technique for image processing, and threshold segmentation is crucial in image segmentation [19], how to determine the boundary is the key to multi-level thresholding, we divide the image into many regions [20]. As an example, if we want to determine s thresholds of the original image [21], we need to divide the pixels of the image into (s + 1) categories [22]. Reference 35 also introduces a concept of cross-entropy, where the theoretical distance between two probability distributions is measured by cross-entropy.

Finally, the tumbleweed algorithm is applied to image segmentation [23]; image segmentation technique is a classical problem in the field of computer vision [24]. It is often mentioned in the current target detection and pattern recognition [25, 26]. Through the pixel information of an image [27], the image is divided into different regions by choosing different thresholds, for example, double thresholding is to divide the pixels of a threshold into different two types of information.

The rest of the paper is organized as follows. Section 17.2 describes the tumbleweed algorithm. Section 17.3 conducts the experiments as well as analyzes the experimental results. Section 17.4 gives the conclusion [28].

17.2 Tumbleweed Algorithm

This section introduces the tumbleweed algorithm and some of its public notices. The tumbleweed algorithm simulates the growth of young shoots and mature reproduction. For ease of description, the current space is denoted as R. There are n tumbleweed seedlings in R. Each tumbleweed seedling is denoted by X_i(i = 1,2,...,N), the fitness is denoted by f(X_i), and the most thriving tumbleweed grows is denoted by X_{gbest}. The pseudo-code of the algorithm is shown in Algorithm 17.1, and the two phases of TA are described below:

17.2.1 Seedling Growth Stage

$$Factor1 = r_1 * \left(X_{gbest} - X^i_{old}\right) \quad (17.1)$$

$$Factor2 = \frac{r_2}{\frac{fit(X^i_{old})}{sum(fit(X_{old}))} + 1} \tag{17.2}$$

Algorithm 17.1: Tumbleweed Algorithm

```
Require: f (x): objective function; dim: dimension;
gc: growth cycle; Max iter: maximum number of iterations;
Ensure: optimal X_best and optimal value f * (X_best);
    Initialize the population X;
    repeat
        Calculate the fitness of each individual;
    Update X_best and f * (X_best);
        if mod(iter, gc)< gc/2 then
        compute Factor1 using Eq.1;
        compute Factor2 using Eq.2;
        update r1 using Eq.3;
            compute X_new using Eq.4;
        else
            compute X_new using Eq.6;
        end if
    until (iter<Max_iter)
```

The factor in Eqs. 17.1 and 17.2 represents the two influences caused by nature on the growth of tumbleweed. Factor1 represents the influence of mature tumbleweed on the seedling, and Factor2 represents the influence of natural factors such as air and soil on the seedling. The parameter r1 is the influence of mature tumbleweeds on seedlings, r_1 will decrease to 0 with eq. 17.3, and r_2 is a random number between 0 and 0.1.

$$r_1 = t * \left(1 - \frac{iter}{\text{Max}iter}\right) \tag{17.3}$$

Finally, the first stage updates the formula to Eq. 17.4.

$$X^i_{new} = X^i_{old} + Factor1 + Factor2 \tag{17.4}$$

17.2.2 Mature Seeding Stage

$$x = \frac{\mu H}{F} \tag{17.5}$$

According to the literature [29], the state of seed propagation should be as shown in Eq. 17.5. x and H represent the distance that the seed can be propagated and the height of the plant at the time of sowing, respectively, and sometimes H and F are just two constants. In TA, H and F are defined as the current iteration number iter and the maximum iteration number *Max_iter*. The mature seeding stage equations are summarized as follows:

$$X_{new}^{i} = X_{gbest} + V_i * \frac{iter}{\text{Max}iter} \tag{17.6}$$

where V_i is the random number in [lb, ub].

17.3 Application in Multilevel Image Segmentation Problem

In this section, the experiment uses four original images as shown in Figs. 17.1, 17.2, 17.3, 17.4, 17.5, 17.6, 17.7, 17.8, 17.9, and 17.10. In the experiment, the number of thresholds for the three channels (RGB) is set to 5, 8, and 11, respectively, and the experimental effect of TA on image segmentation is compared with the ROA and PSO algorithms. PSO is the particle swarm optimization algorithm, which was inspired by a flock of birds for foraging, and has been improved and discussed by a large number of researchers over the years. The proposal of PSO also provided inspiration for various meta-heuristic algorithms that came later. ROA is the rhododendron algorithm, which is inspired by the plant rhododendron. In order to be fair, each image is run 50 times using three algorithms respectively, with 100

Fig. 17.1 Original image

Fig. 17.2 TA (segmented images for 5 thresholding values)

Fig. 17.3 TA (segmented images for 8 thresholding values)

Img1 Img2 Img3 Img4

Fig. 17.4 TA (segmented images for 11 thresholding values)

Img1 Img2 Img3 Img4

Fig. 17.5 PSO (segmented images for 5 thresholding values)

Img1 Img2 Img3 Img4

Fig. 17.6 PSO (segmented images for 8 thresholding values)

Img1 Img2 Img3 Img4

Fig. 17.7 PSO (segmented images for 11 thresholding values)

iterations and an initialized population size of 50, the segmentation of the image using different algorithms is shown in Figs. 17.1, 17.2, 17.3, 17.4, 17.5, 17.6, 17.7, 17.8, 17.9, and 17.10. We found that the larger the number of thresholds the higher the quality of the image segmentation will be [30]. The experiment also evaluated the segmented image using three evaluation metrics (PSNR, SSIM, FSIM), the larger the value of the three metrics represents the higher quality of the image. The results of the experiment are shown Table 17.1 and it is clearly seen that when the threshold is high the values of the three metrics also follow Tables 17.2 and 17.3.

Img1 Img2 Img3 Img4

Fig. 17.8 ROA (segmented images for 5 thresholding values)

Img1 Img2 Img3 Img4

Fig. 17.9 ROA (segmented images for 8 thresholding values)

Img1 Img2 Img3 Img4

Fig. 17.10 ROA (segmented images for 11 thresholding values)

Table 17.1 PSNR

	TA	ROA	PSO
Img1 (5 thresholds)	32.7979	32.7948	32.7948
Img1 (8 thresholds)	35.7896	35.8039	35.7272
Img1 (11 thresholds)	38.6286	38.3100	38.5933
Img2 (5 thresholds)	32.8588	32.8244	32.8244
Img2 (8 thresholds)	36.252	36.1510	36.1892
Img2 (11 thresholds)	38.5829	39.0227	39.1915
Img3 (5 thresholds)	33.3715	33.3504	33.3504
Img3 (8 thresholds)	36.1635	36.6057	36.6608
Img3 (11 thresholds)	38.9694	38.7896	39.2691
Img4 (5 thresholds)	31.5791	31.5250	31.5250
Img4 (8 thresholds)	35.1767	35.1661	35.1331
Img4 (11 thresholds)	38.0007	37.7331	37.5826

17.4 Conclusion

In this paper, we use a new meta-heuristic algorithm TA. TA provides a full rendition of the growth and reproduction stages of tumbleweeds, which improves the algorithmic capability to a relatively high degree. In addition, we apply TA to the multistage image segmentation problem. Three evaluation metrics, PSNR (Peak

Table 17.2 SSIM

	TA	ROA	PSO
Img1 (5 thresholds)	0.8892	0.8880	0.8917
Img1 (8 thresholds)	0.9211	0.9216	0.9206
Img1 (11 thresholds)	0.9410	0.9427	0.9427
Img2 (5 thresholds)	0.9281	0.9275	0.9275
Img2 (8 thresholds)	0.9654	0.9536	0.9558
Img2 (11 thresholds)	0.9662	0.9662	0.9682
Img3 (5 thresholds)	0.8732	0.8730	0.8731
Img3 (8 thresholds)	0.9374	0.9370	0.9377
Img3 (11 thresholds)	0.9511	0.9476	0.9580
Img4 (5 thresholds)	0.9226	0.9219	0.9219
Img4 (8 thresholds)	0.9619	0.9615	0.9597
Img4 (11 thresholds)	0.9723	0.9752	0.9745

Table 17.3 FSIM

	TA	ROA	PSO
Img1 (5 thresholds)	0.9032	0.9018	0.9031
Img1 (8 thresholds)	0.9491	0.9478	0.9474
Img1 (11 thresholds)	0.9682	0.9706	0.9704
Img2 (5 thresholds)	0.9521	0.9503	0.9503
Img2 (8 thresholds)	0.9762	0.9762	0.9766
Img2 (11 thresholds)	0.9852	0.9863	0.9858
Img3 (5 thresholds)	0.9146	0.9149	0.9151
Img3 (8 thresholds)	0.9656	0.9671	0.9677
Img3 (11 thresholds)	0.9839	0.9782	0.9826
Img4 (5 thresholds)	0.9430	0.9418	0.9418
Img4 (8 thresholds)	0.9757	0.9757	0.9746
Img4 (11 thresholds)	0.9892	0.9881	0.9861

Signal-to-Noise Ratio), SSIM (Structural SIMilarity), and FSIM (Feature Similarity Index Measure), were also used to evaluate the quality of the segmented image. PSNR is simple to compute, if the input is a color image, the PSNR values for each of the three RGB channels are computed separately, and then averaged. SSIM measures three aspects of the image data, namely luminance, structure, and contrast. A window of size m*n slides continuously over the image, calculating values at each position. Finally, the average of these values is taken as the final result. Fsim on the other hand uses feature similarity to assess the quality of the image and it uses two features pc and gm to complement it. The experimental results show that TA has a good performance for image segmentation, and in our future work, we will optimize the TA algorithm by adding parallel or hybrid strategies to further enhance the algorithm's capability and apply it to other practical problems.

References

1. Price, K.V.: Differential evolution. In: Handbook of Optimization: from Classical to Modern Approach, pp. 187–214. Springer (2013)
2. Yang, Q., Chu, S.C., Liang, A., Pan, J.S.: Tumble- weed algorithm and its application for solving location problem of logistics distribution center. In: Genetic and Evolutionary Computing: Proceedings of the Fourteenth International Conference on Genetic and Evolutionary Computing, pp. 641–652. Springer (2022)
3. Kennedy, J., Eberhart, R.: Particle swarm optimization. In: Proceedings of ICNN'95-International Conference on Neural Networks, vol. 4, pp. 1942–1948. IEEE (1995)
4. Wu, J., Xu, M., Liu, F.-F., Huang, M., Ma, L., Lu, Z.-M.: Solar wireless sensor network routing algorithm based on multi-objective particle swarm optimization. J. Inf. Hiding Multim. Signal Process. **12**(1), 1–11 (2021)
5. Karaboga, D., Gorkemli, B., Ozturk, C., Karaboga, N.: A comprehensive survey: artificial bee colony (abc) algorithm and applications. Artif. Intell. Rev. **42**, 21–57 (2014)
6. Chu, S.C., Tsai, P.-W., Pan, J.-S.: Cat swarm optimization. In: PRICAI 2006: Trends in Artificial Intelligence: 9th Pacific Rim International Conference on Artificial Intelligence, pp. 854–858. Springer (2006)
7. Arora, S., Singh, S.: Butterfly optimization algorithm: a novel approach for global optimization. Soft. Comput. **23**, 715–734 (2019)
8. Yang, X.-S., Deb, S.: Cuckoo search: recent advances and applications. Neural Comput. & Applic. **24**, 169–174 (2014)
9. Abualigah, L., Yousri, D., Abd Elaziz, M., Ewees, A.A., et al.: Aquila optimizer: a novel metaheuristic optimization algorithm. Comput. Ind. Eng. **157**, 107250 (2021)
10. Askari, Q., Younas, I., Saeed, M.: Political optimizer: a novel socio-inspired meta-heuristic for global optimization. Knowl.-Based Syst. **195**, 105709 (2020)
11. Pan, J.-S., Meng, Z., Xu, H., Li, X.: Quasi-affine transformation evolution (quatre) algorithm: a new simple and accurate structure for global optimization. In: Trends in Applied Knowledge-Based Systems and Data Science: 29th International Conference on Industrial Engineering and Other Applications of Applied Intelligent Systems, IEA/AIE 2016, pp. 657–667. Springer (2016)
12. Pan, J.-S., Zhang, L.-G., Wang, R.-B., Snášel, V., Chu, S.-C.: Gannet optimization algorithm: a new metaheuristic algorithm for solving engineering optimization problems. Math. Comput. Simul. **202**, 343–373 (2022)
13. Duan, H., Qiao, P.: Pigeon-inspired optimization: a new swarm intel-ligence optimizer for air robot path planning. Int. J. Intell. Comput. Cybernet. **7**(1), 24–37 (2014)
14. Mirjalili, S., Lewis, A.: The whale optimization algorithm. Adv. Eng. Softw. **95**, 51–67 (2016)
15. Chiu, M.-S., Lin, C.-C., Cheng, C.-S., Yeh, M.-T., Chung, Y.-N., Hsu, C.-H.: Applying image processing algorithm to dynamic face detection. J. Inf. Hiding Multim. Signal Process. **12**(4), 207–216 (2021)
16. Hu, P., Chu, S.-C., Snášel, V., Pan, J.-S.: Binary equilibrium optimizer algorithm. J. Netw. Intell. (2022)
17. Wang, H., Liang, M., Sun, C., Zhang, G., Xie, L.: Multiplestrategy learning particle swarm optimization for large-scale optimization problems. Complex Intell. Syst. **7**, 1–16 (2021)
18. Pan, J.-S., Shi, H.-J., Chu, S.-C., Hu, P., Shehadeh, H.A.: Parallel binary rafflesia optimization algorithm and its application in feature selection problem. Symmetry. **15**(5), 1073 (2023)
19. Yang, H., Pu, Z., Guo, Z., Zhang, A., Xiong, S.: A study of metal/die interfacial heat transfer behavior of vacuum die cast pure copper. China Foundry. **17**, 206–211 (2020)
20. Wang, X., Chu, S.-C., Snášel, V., Shehadeh, H.A., Pan, J.-S.: Five phases algorithm: a novel meta-heuristic algorithm and its application on economic load dispatch problem. J. Internet Technol. **24**(4), 837–848 (2023)
21. Fan, K., Liang, J., Li, F., Qiu, P.: Cnn based no-reference hdr image quality assessment. Chin. J. Electron. **30**(2), 282–288 (2021)

22. Gong, C., Liu, J., Niu, Y.: Intracranial epileptic seizures detection based on an optimized neural network classifier. Chin. J. Electron. **30**(3), 419–425 (2021)
23. Yu, X., Wu, L., Xie, M., Wang, Y., Xu, L., Lu, H., Xu, C.: Smoothing neural network for non-lipschitz optimization with linear inequality constraints. Chin. J. Electron. **30**(4), 634–643 (2021)
24. Min-Allah, N., Nagy, N., Aljabri, M., Alkharraa, M., Alghamdi, M.A.D., Sabri, R., Alshaikh, R.: Quantum image steganography schemes for data hiding: a survey. Appl. Sci. **12**(20), 10294 (2022)
25. Khan, M., Rasheed, A.: A secure controlled quantum image steganog-raphy scheme based on the multi-channel effective quantum image representation model. Quantum Inf. Process. **22**(7), 268 (2023)
26. Varghese, F., Sasikala, P.: Secure data transmission using optimized cryptog-raphy and steganography using syndrome-trellis coding. Wirel. Pers. Commun. **130**(1), 551–578 (2023)
27. Palsodkar, P., Khandare, A., Raut, A., Giradkar, V., Dekate, S., Rokade, R.: Empirical analysis of quantum compution using qiskit framework. In: 2023 9th International Conference on Advanced Computing and Communication Systems (ICACCS), vol. 1, pp. 467–472. IEEE (2023)
28. Yue, L., Hu, P., Chu, S.-C., Pan, J.-S.: Multi-objective gray wolf optimizer with cost-sensitive feature selection for predicting students' academic performance in college english. Mathematics. **11**(15), 3396 (2023)
29. Nathan, R., Katul, G.G., Horn, H.S., Thomas, S.M., Oren, R., Avissar, R., Pacala, S.W., Levin, S.A.: Mechanisms of long-distance dispersal of seeds by wind. Nature. **418**(6896), 409–413 (2002)
30. Nguyen, T.-T., Nguyen, T.-D., Ngo, T.-G., Nguyen, V.-T.: An optimal thresholds for segmenting medical images using improved swarm algorithm. J. Inf. Hiding Multim. Signal Process. **13**(1), 12–21 (2022)

Chapter 18
DiffTri-Network: An Architecture for Medical Image Segmentation Using Non-expert Annotations

Weimin Zheng, Chengyu Jiang, Shicheng Guo, Jingyu Wang, and Shangkun Liu

Abstract Convolutional neural networks (CNNs) have achieved remarkable performance in various medical image segmentation tasks. However, the success of CNNs heavily relies on a large amount of accurately annotated training samples, which are challenging and expensive to obtain in medical image analysis. In clinical settings, the task of annotating lesions is gradually being performed by trained non-expert personnel, making data more accessible. Leveraging non-expert annotated data to extract reliable information for lesion segmentation, addressing the challenge of obtaining expert annotations for segmentation tasks by utilizing these labels. In this paper, we propose an architecture called DiffTri-Network for medical image segmentation tasks using non-expert annotations, aiming to alleviate the challenges of data unavailability. Specifically, our architecture employs network models of different depths that fuse their respective extracted reliable features, guiding the network's learning process and updating their parameters respectively. These three networks achieve collaborative optimization through this co-learning approach. In order to simulate non-expert annotated medical images in real-life scenarios, we mixed the benign and malignant tumor labels in the BUSI dataset and applied dilation, scaling, and jitter operations. To ensure the objectivity of the data, the data generated from these three operations were randomly partitioned. We compare our method on this dataset with fully supervised segmentation methods. Experimental results show that our method is capable of extracting more image features from such data and achieves excellent segmentation performance.

W. Zheng · C. Jiang · S. Guo · J. Wang · S. Liu (✉)
Shandong University of Science and Technology, Qingdao, China
e-mail: guoshicheng@sdust.edu.cn

S.-P. Tseng et al. (eds.), *Advances in Intelligent Information Hiding and Multimedia Signal Processing, Volume 2*, Smart Innovation, Systems and Technologies 416, https://doi.org/10.1007/978-981-97-8760-9_18

18.1 Introduction

Lesion segmentation is one of the critical issues in medical image processing. These lesion annotations are more accessible to obtain but only provide rough lesion boundaries, lacking detailed accuracy [1]. These annotations can be considered as a source of imperfect labels. To address the problem of complex acquisition of expert-annotated labels, we can train the network using defective labels to obtain more accurate lesion segmentation results [2]. Due to the significant noise interference, which can affect the performance of the network model, it is necessary to design unique training architectures to mitigate the impact of noise on neural network training.

Remembering foreground pixels with incorrect labels may lead to erroneous judgments of background pixels, and this accumulation of errors can decrease the accuracy of foreground pixel predictions. Therefore, mitigating the impact of erroneously labeled samples on the model is crucial. Arpit et al. [3] proposed that deep networks perform significantly better when the sample is as clean as possible. Han et al. [1] introduced joint training of two networks, where each network is alternately updated, ensuring the performance of the network model. In a CNN's relatively late training stages, samples with relatively lower losses are treated as clean data to help the model fit the data better. Also, meta-learning has been introduced into medical image analysis [4], where pixel importance is reweighted based on loss analysis to reduce the influence of edge noise. For medical image analysis tasks, Xue et al. [5] proposed a sample reweighting framework that removes samples with high loss during network training and reweights the retained samples within each mini-batch. Lehtinen et al. [6] proposed a method for learning from noisy samples, which can be applied to MR image reconstruction. Although these methods have shown exemplary performance in image classification and object detection tasks, they cannot be directly applied to image segmentation tasks [7].

In image segmentation tasks, noisy images can mislead neural networks to extract global and local features, resulting in poor segmentation performance [8]. To address this issue, Zhu et al. [9] proposed a label quality assessment strategy that automatically evaluates the quality of labels and preserves important pixel information in the labels. Another approach is to generate weight maps to guide the importance of pixels, reducing the influence of noisy pixels. Tri-Network [10], introduced in 2020, involves three network models that learn collaboratively and adaptively select feature information based on standard pixels among network predictions. Based on these works, medical image segmentation methods dealing with noisy labels can be summarized into three categories. The first category involves pixel-wise label updates through label smoothing regularization [11]. The second category models the distribution of noisy labels using confusion matrices or noise transition matrices (NTMs) [12] and then performs pixel-level loss correction. The third category employs strategies such as pixel resampling and reweighting to focus the neural network on learning reliable pixels. In addition to these methods, intelligent computing algorithms are also applied to noise label learning, optimizing segmentation performance through algorithms [13–15].

A new learning framework has been proposed to address the issue mentioned above, utilizing imperfectly labeled data to reduce the burden of high-quality annotation work and mitigate the negative impact of noise labels [16–18]. Inspired by the collaborative learning and Tri-Network, a novel training architecture is introduced. Building upon the Tri-Network architecture, which employs three network models with the same architecture but different parameters, the models are trained simultaneously and complement each other's predictions to ensure effective feature extraction for segmentation. Following the principles of the Tri-Network architecture, we embed networks of different depths and train them concurrently while addressing feature complementarity [15, 19]. This approach supplements the missing pixels in each network's feature extraction using the additional features extracted by other networks. Furthermore, we streamline the subsequent feature supplementation process to achieve a lightweight architecture. To ensure the objectivity of the data, the data generated from these three operations were randomly partitioned. We evaluate our method on this dataset against several state-of-the-art fully supervised segmentation methods.

18.2 Method

18.2.1 Overview

The architecture we propose, called DiffTri-Network, is shown in Fig. 18.1. The main idea behind this architecture is to simultaneously train three different depth network models, each of which extracts essential and reliable information from non-expert annotations. Given a batch of input data, they are inputted into the three networks separately, generating three different prediction maps. Next, for each pair of networks (e.g., Network 1 and Network 2), the other prediction maps undergo reliable pixel-selection and feature fusion, resulting in a mask (e.g., Mask 12) representing the selected reliable pixels. Subsequently, we compute the loss between the ground truth labels and the mask, take the weighted average of the three loss values to guide the parameter updates of the three networks. In each training iteration, each network repeats the same operation. After training, the test data is inputted into the most profound network model, producing the final mask. Figure 18.1 depicts the overall architecture of DiffTri-Network, which is used for medical image segmentation tasks based on non-expert annotations.

18.2.2 Feature Extraction

We generate different predictions for the same data during the network training process using various networks. We select reliable pixels from the different predictions and generate a mask as reliable samples. These reliable samples from each group

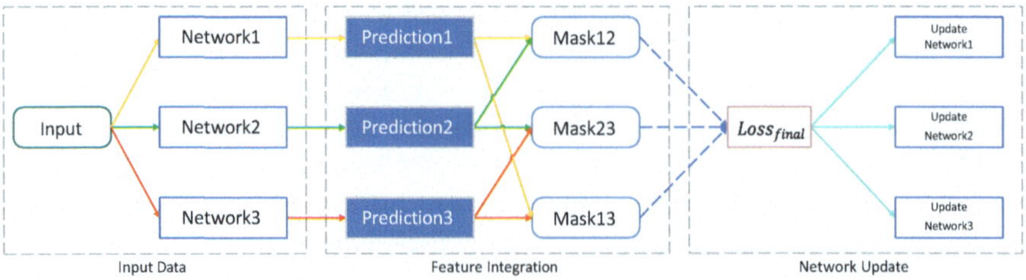

Fig. 18.1 The training architecture of DiffTri-Network is used for medical image segmentation tasks based on non-expert annotations. We show an overview of the training process for a single iteration

are then used for feature fusion. The fused mask is used to guide the training process of each group of networks, enabling the networks to focus more on these valuable pixels and reduce the impact of noisy labels.

Feature Amalgamation For each batch of data, the network ensemble generates two pixel-level prediction maps. These prediction maps represent the segmentation results of each network on the input image. Due to each network's unique architectures and learning capabilities, they focus on different aspects of feature extraction. To leverage this characteristic, we perform pixel-wise multiplication of the prediction maps from each network pair within the ensemble, merging the feature information between the two networks. This process results in three new pixel-level prediction maps. After feature fusion, we apply thresholding and connected component analysis operations to the fused prediction maps. This fusion approach reduces potential inaccuracies in individual networks and enhances the multi-faceted representation of feature information.

Parameter Update After feature fusion, the fused predicted feature map is compared with the corresponding label using the Dice loss calculation. The three individual losses are then weighted and averaged to obtain the final composite loss for backpropagation. The formulas for the Dice loss function and the final loss function are shown as Eqs. (18.1) and (18.2):

$$loss_j = DicsLoss = 1 - \frac{2\sum_{i=1}^{N} y_i \hat{y}_i}{\sum_{i=1}^{N} y_i + \sum_{i=1}^{N} \hat{y}_i} \tag{18.1}$$

$$Loss_{final} = \frac{1}{N}\sum_{j=1}^{N} loss_j \tag{18.2}$$

where j is the *j-th* network model.

18.2.3 Technical Details

This framework is implemented using PyTorch on the Nvidia RTX 3090 GPU. We have three different networks in our framework with varying levels of complexity and segmentation performance. We primarily chose U-Net, U-Net++, and DSCAU-Net as the backbone architectures. These three networks complement each other's features during the training process. The entire framework uses the Adam optimizer with a momentum of 0.9. The networks are trained with a learning rate of 0.01 for 500 epochs or until the model converges. We set the batch size to 4 so that all three networks can be trained simultaneously on a single GPU. The input images are all resized to 256 × 256 pixels. The predicted images generated by each network are compared with the ground truth labels using the dice loss calculation. The three losses are then weighted and averaged to compute the final loss for back-propagation.

18.3 Experiment

18.3.1 Dataset

We evaluate our proposed architecture using the open-source dataset BUSI [20]. The BUSI dataset is widely used in breast ultrasound image processing and contains 600 cases of breast ultrasound of women aged 25–75. The dataset consists of 780 images with a size of 500 × 500, including 437 benign cases, 210 benign cases, and 133 normal cases. To more carefully evaluate the performance of our vector for malignant and benign tumor segmentation, we train and test on the BUSI dataset for benign and malignant tumors, respectively. In the experiment, we selected 452 images as the training set, 128 as the validation set, and the remaining as the test set. The specific allocation is shown in Table 18.1. Among them, the dilated label data accounts for 30% of the training set, the eroded label data accounts for 40%, and the jittered label data accounts for 30%. The partitioning of non-expert annotated labels in the training set is randomly generated to ensure the simulation's objectivity. The validation set and the test set consist of original labels.

Table 18.1 Illustration of the BUSI dataset used in this work

Datasets	Train	Valid	Test	Total
Benign	313	79	45	437
Malignant	139	49	22	210

18.3.2 Experiment Configurations and Evaluation Metrics

All experiments are performed in parallel on a single RTX 3090 GPU using the deep learning PyTorch framework. We use the proposed DiffTri-Network as the backbone network, and use this network to generate a mask for segmentation. In the network training, three networks are set to train data. To quantitatively compare the segmentation results of different methods on breast tumors, we used several evaluation indicators such as DSC, Jaccard, Accuracy, Precision, and Recall.

18.3.3 Model Performance Analysis

We compared the segmentation results of our proposed DiffTri-Network architecture with existing fully supervised segmentation methods on the BUSI dataset. The quantitative results are shown in Table 18.2. On the mixed dataset, our approach demonstrates significant advantages over other methods in DSC, Jaccard, and Recall metrics. The table shows that our proposed DiffTri-Network method improves by 2% and 1% in DSC and Jaccard, and enhances Recall by 1%, compared to the Tri-Network method. The proposed architecture simplifies the feature fusion and network updating components compared to the Tri-Network, which results in a slight decrease in Accuracy and Precision. However, the reduction in these two metrics is minimal, and the performance remains good. This indicates that our proposed DiffTri-Network architecture outperforms some single-network segmentation methods. In addition to the version improvement, our approach also reduces computational complexity and processing time compared to the Tri-Network. The visualization of the segmentation results of the DiffTri-Network on the BUSI dataset is depicted in Fig. 18.2.

Table 18.2 Quantitative comparison of our proposed DiffTri-Network on BUSI dataset, with the existing segmentation methods

Method	Fully Supervised?	DSC (%)	Jaccard (%)	Accuracy (%)	Precision (%)	Recall (%)
U-Net [21]	✓	59.60	47.34	92.77	55.34	74.54
U-Net++ [22]	✓	58.24	46.35	92.56	59.43	69.03
ResU-Net++ [23]	✓	43.99	33.02	91.49	43.56	57.30
AttentionU-Net [24]	✓	58.90	46.39	93.17	55.60	75.81
DSCU-Net [25]	✓	61.93	49.24	92.56	60.11	74.74
DSCAU-Net [26]	✓	58.20	47.15	92.68	56.18	72.39
Tri-Network [10]	×	61.82	49.69	93.59	62.22	72.56
DiffTri-Network(U-Net)	×	**60.53**	**48.83**	**93.38**	**60.25**	**72.76**
DiffTri-Network(U-Net++)	×	**63.79**	**50.83**	**92.13**	**58.40**	**81.17**
DiffTri-Network(DSCAU-Net)	×	**62.55**	**51.17**	**93.21**	**59.22**	**76.94**

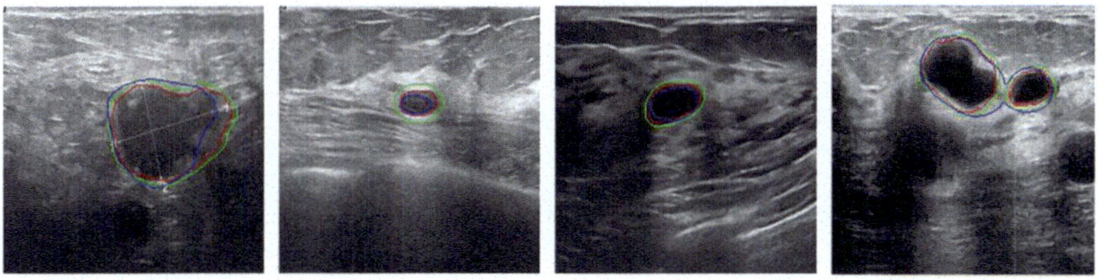

Fig. 18.2 Visual examples on the BUSI dataset. The red, green, and blue colors denote the ground truth annotation, DiffTri-network, and the noisy annotation. Our method generates more accurate contours than non-expert annotation

We evaluated three networks trained using the proposed architecture to validate our approach and demonstrate the improved segmentation performance. We compared the segmentation results of these networks with those obtained from individually trained networks, and the comparative results are summarized in Table 18.2. It can be observed that the networks trained with the DiffTri-Network architecture exhibit improved segmentation performance. This confirms that our proposed method enhances the network's performance, enabling it to achieve superior segmentation accuracy and precision.

18.4 Conclusion

In this work, we propose the DiffTri-Network architecture that leverages the complementary learning abilities of different networks to address the challenge of accurate medical image segmentation using noisy labels. Our proposed architecture simultaneously trains three networks with varying complexities. The predictions generated by each network are fused and enriched to guide the network's learning process. The entire architecture is designed to focus on collaborative learning, leveraging feature fusion to connect the three networks and facilitate information exchange between the learned features, thereby improving segmentation accuracy. We validate the effectiveness of our proposed framework on a non-expert annotated dataset. Experimental results demonstrate that our method enhances the network's ability to extract features from noisy images and achieves excellent segmentation performance.

Conflicts of Interest The authors declare no conflict of interest.

Funding Statement The authors acknowledge the funding of the National Natural Science Foundation of China, Grant No. 11974373.

Data Availability The data used to support the findings of this study are included within the article.

References

1. Han, B., Yao, Q., Yu, X., Niu, G., Xu, M., Hu, W., Tsang, I., Sugiyama, M.: Co-teaching: robust training of deep neural networks with extremely noisy labels. Adv. Neural Inf. Proces. Syst. **31** (2018)
2. Ching, T., Himmelstein, D.S., Beaulieu-Jones, B.K., Kalinin, A.A., Do, B.T., Way, G.P., Ferrero, E., Agapow, P.-M., Zietz, M., Hoffman, M.M., et al.: Opportunities and obstacles for deep learning in biology and medicine. J. R. Soc. Interface. **15**(141), 20170387 (2018)
3. Arpit, D., Jastrzębski, S., Ballas, N., Krueger, D., Bengio, E., Kanwal, M.S., Maharaj, T., Fischer, A., Courville, A., Bengio, Y., et al.: A closer look at memorization in deep networks. In: International Conference on Machine Learning, pp. 233–242. PMLR (2017)
4. Wang, J., Zhou, S., Fang, C., Wang, L., Wang, J.: Meta corrupted pixels mining for medical image segmentation. In: Medical Image Computing and Computer Assisted Intervention–MICCAI 2020: 23rd International Conference, Part I, vol. 23, pp. 335–345. Springer (2020)
5. Xue, C., Dou, Q., Shi, X., Chen, H., Heng, P.-A.: Robust learning at noisy labeled medical images: applied to skin lesion classification. In: 2019 IEEE 16th International Symposium on Biomedical Imaging (ISBI 2019), pp. 1280–1283. IEEE (2019)
6. Lehtinen, J., Munkberg, J., Hasselgren, J., Laine, S., Karras, T., Aittala, M., Aila, T.: Noise2noise: Learning image restoration without clean data. arXiv preprint arXiv:1803.04189 (2018)
7. Zheng, W., Zhou, L., Chai, Q., Xu, J., Liu, S.: Fully automatic analysis of muscle b-mode ultrasound images based on the deep residual shrinkage u-net. Electronics. **11**(7), 1093 (2022)
8. Yang, L., Qi, L., Feng, L., Zhang, W., Shi, Y.: Revisiting weak-to-strong consistency in semi-supervised semantic segmentation. In: Proceedings of the IEEE/CVF Conference on Computer Vision and Pattern Recognition, pp. 7236–7246 (2023)
9. Zhu, X., Chen, J., Zeng, X., Liang, J., Li, C., Liu, S., Behpour, S., Xu, M.: Weakly supervised 3d semantic segmentation using cross-image consensus and inter-voxel affinity relations. In: Proceedings of the IEEE/CVF International Conference on Computer Vision, pp. 2834–2844 (2021)
10. Zhang, T., Yu, L., Hu, N., Lv, S., Gu, S.: Robust medical image segmentation from non-expert annotations with tri-network. In: Medical Image Computing and Computer Assisted Intervention–MICCAI 2020: 23rd International Conference, Part IV, vol. 23, pp. 249–258. Springer (2020)
11. Xu, Z., Lu, D., Wang, Y., Luo, J., Jayender, J., Ma, K., Zheng, Y., Li, X.: Noisy labels are treasure: mean-teacher-assisted confident learning for hepatic vessel segmentation. In: Medical Image Computing and Computer Assisted Intervention–MICCAI 2021: 24th International Conference, Part I, vol. 24, pp. 3–13. Springer (2021)
12. Guo, X., Yang, C., Li, B., Yuan, Y.: Metacorrection: domain-aware meta loss correction for unsupervised domain adaptation in semantic segmentation. In: Proceedings of the IEEE/CVF Conference on Computer Vision and Pattern Recognition, pp. 3927–3936 (2021)
13. Chu, S.-C., Zhuang, Z., Pan, J.-S., Mohamed, A.W., Hu, C.-C.: Enhanced sparseea for large-scale multi-objective feature selection problems. Complex Intell. Syst. **1–23**, 485 (2023)
14. Zhou, J.-L., Chu, S.-C., Tian, A.-Q., Peng, Y.-J., Pan, J.-S.: Intelligent neural network with parallel salp swarm algorithm for power load prediction. J. Internet Technol. **23**(4), 643–657 (2022)
15. Chu, S.-C., Roddick, J.F., Pan, J.-S.: Ant colony system with communication strategies. Inf. Sci. **167**(1–4), 63–76 (2004)
16. Zheng, W., Pang, S., Liu, N., Chai, Q., Xu, L.: A compact snake optimization algorithm in the application of wknn fingerprint localization. Sensors. **23**(14), 6282 (2023)
17. Pan, J.-S., Hu, P., Snášel, V., Chu, S.-C.: A survey on binary metaheuristic algorithms and their engineering applications. Artif. Intell. Rev. **56**(7), 6101–6167 (2023)

18. Yang, Q., Chu, S.-C., Pan, J.-S., Chou, J.-H., Watada, J.: Dynamic multistrategy integrated differential evolution algorithm based on reinforcement learning for optimization problems. Complex Intell. Syst. **1–33**, 1845 (2023)
19. Zhuang, Z., Pan, J.-S., Li, J., Chu, S.-C.: Parallel binary arithmetic optimization algorithm and its application for feature selection. Knowl. Based Syst. **275**, 110640 (2023)
20. Al-Dhabyani, W., Gomaa, M., Khaled, H., Fahmy, A.: Dataset of breast ultrasound images. Data Brief. **28**, 104863 (2020)
21. Ronneberger, O., Fischer, P., Brox, T.: U-net: convolutional networks for biomedical image segmentation. In: Medical Image Computing and Computer-Assisted Intervention–MICCAI 2015: 18th International conference, Part III, vol. 18, pp. 234–241. Springer (2015)
22. Zhou, Z., Rahman Siddiquee, M.M., Tajbakhsh, N., Liang, J.: Unet++: a nested u-net architecture for medical image segmentation. In: Deep Learning in Medical Image Analysis and Multimodal Learning for Clinical Decision Support: 4th International Workshop, DLMIA 2018, and 8th International Workshop, ML-CDS 2018, Held in Conjunction with MICCAI 2018, vol. 4, pp. 3–11. Springer (2018)
23. Jha, D., Smedsrud, P.H., Riegler, M.A., Johansen, D., De Lange, T., Halvorsen, P., Johansen, H.D.: Resunet++: an advanced architecture for medical image segmentation. In: 2019 IEEE International Symposium on Multimedia (ISM), pp. 225–2255. IEEE (2019)
24. Oktay, O., Schlemper, J., Folgoc, L.L., Lee, M., Heinrich, M., Misawa, K., Mori, K., McDonagh, S., Hammerla, N.Y., Kainz, B., et al.: Attention u-net: Learning where to look for the pancreas. arXiv preprint arXiv:1804.03999 (2018)
25. Xu, L., Ming, D., Du, T., Chen, Y., Dong, D., Zhou, C.: Delineation of cultivated land parcels based on deep convolutional networks and geographical thematic scene division of remotely sensed images. Comput. Electron. Agric. **192**, 106611 (2022)
26. Xu, Q., Ma, Z., Na, H., Duan, W.: Dcsau-net: a deeper and more compact split-attention u-net for medical image segmentation. Comput. Biol. Med. **154**, 106626 (2023)

Chapter 19
Mechanism Design of Short-Distance Food Transmission Robot

Yi-Xian Gu, Yi-Jui Chiu, and Meng Li

Abstract Mechanical transportation provides great convenience to life, and this paper takes the short-distance conveying vehicle as the object of study. Firstly, the model of the short-distance conveyor cart was created by SolidWorks modeling software. Next, the whole vehicle was meshed with the finite element software Workbench, and the stress-strain under different boundary conditions was analyzed qualitatively and quantitatively. The finite element simulation results show that the vibration frequency and stress magnitude of the designed trolley do not exceed the permissible values under various working conditions and meet the requirements of use.

19.1 Introduction

Mechanization is gradually entering the life of the public. Chen et al. [1] used a design and experimental verification approach to study the cost, energy efficiency, safety, mobility, and environmental effects of electric vehicles suitable for short-haul rural bulk cargo transportation, and verified the structural strength reliability and operational performance of the whole vehicle through prototype tests. Qi et al. [2] used field investigation, simulation analysis, and optimal design to study the performance and fruit damage of the inter-row transporter damping system when traveling on mountain orchard roads, and proposed an improvement scheme through optimal design to improve the damping performance and stability of the inter-row transporter. Qi et al. [3] used modeling analysis, control strategy design, and simulation analysis to investigate an ideal model of a vehicle combining transverse swing moment control with a four-wheel steering system, and designed an optimal controller, and evaluated the performance of the control strategy through simulation

Y.-X. Gu · Y.-J. Chiu (✉) · M. Li
School of Mechanical and Automotive Engineering, Xiamen University of Technology, Xiamen, China
e-mail: chiuyijui@xmut.edu.cn

S.-P. Tseng et al. (eds.), *Advances in Intelligent Information Hiding and Multimedia Signal Processing, Volume 2*, Smart Innovation, Systems and Technologies 416, https://doi.org/10.1007/978-981-97-8760-9_19

analysis. The results show that the control strategy can effectively control both the steering lateral deflection angle and the transverse swing angular velocity of the vehicle. Wang et al. [4] designed a monorail transporter including a four-wheeled mobile body and a guiding mechanism. The simulation results show that the speed of the transporter oscillates and gradually converges to a steady state when the friction coefficient is high. Lin et al. [5] used a computer simulation study and experimental validation method to study mainly the steering characteristics of four-wheeled vehicles and built a whole vehicle model for analysis. The correctness and applicability of the model were verified through comparative analysis of simulation data and test results. The above literature has carried out certain modeling analysis studies for the structure of short-distance transport vehicles, and the models established have certain reference significance.

Wang et al. [6] evaluated the bending stiffness and modalities of the aluminum alloy body by means of finite element modeling and simulation analysis, and introduced the concept of generalized structural stiffness by using an overall structural assessment method and model improvement by identifying force transfer paths. Ding et al. [7] use parametric surface modeling technique to realize the whole vehicle mapping mesh division, which is easy to adjust the finite element model. Thin-shell and thick-shell cells are used to simulate the car body and frame, and the finite element calculation model with full-plate and shell cells is established. According to the maximum stress region, an improvement scheme is proposed to reduce the maximum stress. Su et al. [8] investigate the whole-vehicle modeling techniques for side impacts and perform crash simulations using moving barrier modeling techniques. The simulation calculation method is applied to analyze the side impact, including the calculation of structural deformation energy, deformation volume and deformation velocity, and other characteristics. Gang et al. [9] propose a finite element modeling method based on the combination of load equivalence method and auxiliary constraint method for describing the super-static nonlinear coupled load-carrying characteristics of the air spring suspension system. By statically balancing the front and rear axles of the whole vehicle, an analysis method of the super-static nonlinear coupled load-bearing characteristics of the system was established. Zhao et al. [10] model a low-floor engine rear-mounted urban articulated bus and consider the actual connection relationship and rotation of the articulated disc. Through finite element analysis and mathematical modeling, the performance is evaluated under the consideration of sharp cornering conditions, which provides a reference for the optimization of the body skeleton structure.

In summary, this paper presents the design of the structure of a short-distance transport vehicle from a real-life perspective. The finite element method is used to carefully analyze the stress-strain of the short-distance transport vehicle under various operating conditions from several angles. Through continuous improvement, the model is designed to meet engineering applications.

19.2 Modeling Process

An intelligent multi-layer food delivery robot was designed for restaurants using robotics for the problem of changing the center of gravity of the delivery vehicle during transportation and the portability of picking up the food after delivery.

SolidWorks is a 3D computer-aided design software that is widely used in engineering and manufacturing. This software provides users with comprehensive design solutions to support their innovation and optimization in various fields. Considering the structural stability of the shell, the internal fixed connection of the shell and the light weight, etc. The final modeling was adopted using SolidWorks software as shown in Fig. 19.1.

For the study of the transport vehicle, the flow chart is shown in Fig. 19.2. The first is the modeling of the short-distance vegetable transport vehicle. Next is the setup of the finite element model. Finally, the finite element simulation results are obtained.

For the design of the finite element modal analysis architecture, the frequency and modal variation of the conveyor car is analyzed in detail. In the analysis process, 45 steel, a commonly used engineering material with good mechanical and heat treatment properties, was used as the material for the conveyor. The material parameters associated with 45 steel are listed in Table 19.1, which are crucial for the accuracy and reliability of the modal analysis.

In order to solve the numerical solution of a real problem, it is necessary to first discretize the problem into a finite element model and then obtain the numerical solution by solving the matrix equations of the model in FAE. Finite element meshing is a crucial step in finite element numerical simulation analysis, directly impacting the accuracy of subsequent calculations and results. The meshing of the

Fig. 19.1 Intelligent multi-layer food delivery robot structure for restaurants

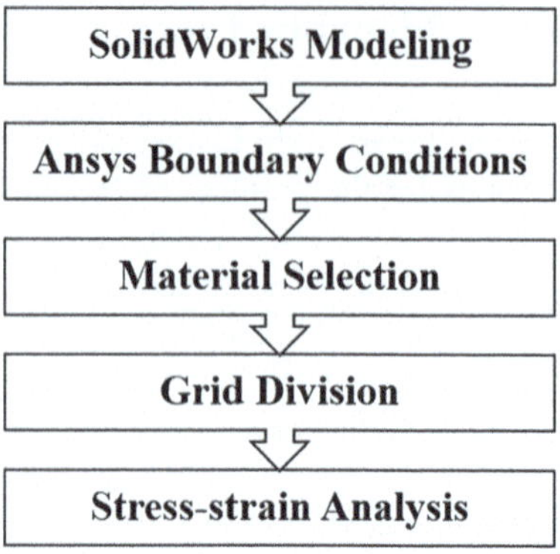

Fig. 19.2 Modeling flow chart

Table 19.1 Material parameters of 45 steel

Property	Numerical value	Unit
Density	7, 850	kg/m^3
Young modulus	2×10^9	Pa
Poisson ratio	0.3	–
Bulk modulus	7.5×10^{10}	Pa
Yield strength	3.55×10^8	Pa
Tangent modulus	1.78×10^8	Pa

Fig. 19.3 Grid division of the transport vehicle

short-distance vegetable truck is shown in Fig. 19.3. The number of generated mesh elements is 71, 173 and the number of nodes is 39, 510.

19.3 Derivation of Basic Formula

19.3.1 Safety Factor Method's Basic Formula

The vehicle safety factor method is a method used to assess the safety of vehicle structures. It determines the reliability and safety of the vehicle structure by calculating the safety factor based on the load requirements and material properties of the vehicle in the design. In the vehicle safety factor method, the design load requirements of the vehicle, including static load and dynamic load, need to be determined first. Then, according to the structural characteristics and material properties of the vehicle, the bearing capacity of each critical part is calculated.

The allowable stress of the vehicle is determined by dividing the ultimate strength by the safety factor.

$$[\sigma] = \frac{\sigma_b}{n} \tag{19.1}$$

The equations of the vehicle structure system are as follows:

$$[M]\left\{\ddot{X}\right\} + [K]\{X\} = 0 \tag{19.2}$$

The inherent frequency of the detuned system is as follows:

$$\overline{\omega_n} = \frac{\omega_n}{\sqrt{EI/\rho AL^4}}, n = 1, 2, 3, \tag{19.3}$$

19.3.2 Basic Formula for Collision Dynamics

Collision dynamics is a subdiscipline of physics that studies the process of object collisions, which involves the transfer and conservation of momentum, energy, and angular momentum. The study of collision dynamics helps to understand the interaction and energy transfer between objects and is important for the interpretation and prediction of collision events in physics, engineering, and other fields.

In a closed system, the total momentum of the object remains constant during the collision. This means that the vector sum of the total momentum of the object before and after the collision remains equal. The Cauchy stress tensor is denoted by σ_{ij}, and the volume force per unit mass is denoted by f_i.

$$\rho \ddot{x}_i = \sigma_{ij} + \rho f_i \tag{19.4}$$

which satisfies the law of conservation of mass at the time of collision, where ρ_0 is the initial density and J represents the deformation gradient determinant, as shown in Eq. (19.5).

$$\rho = J\rho_0 \tag{19.5}$$

In some specific cases of collision, the total kinetic energy of the object remains constant during the collision. This means that the total kinetic energy of the object remains equal before and after the collision. The strain rate tensor is denoted by ε_{ij} and the viscosity resistance is denoted by q.

$$\dot{E} = v s_{ij} \varepsilon_{ij} + (p + q)\,\dot{v} \tag{19.6}$$

19.4 Finite Element Working Condition Analysis

Stress and strain analysis of trolleys under various practical operating conditions. This allows optimizing the design, predicting the system behavior, or performing other analyses. For the purpose of predicting the system behavior, the boundary conditions set for the geometric model entities are shown in Fig. 19.4.

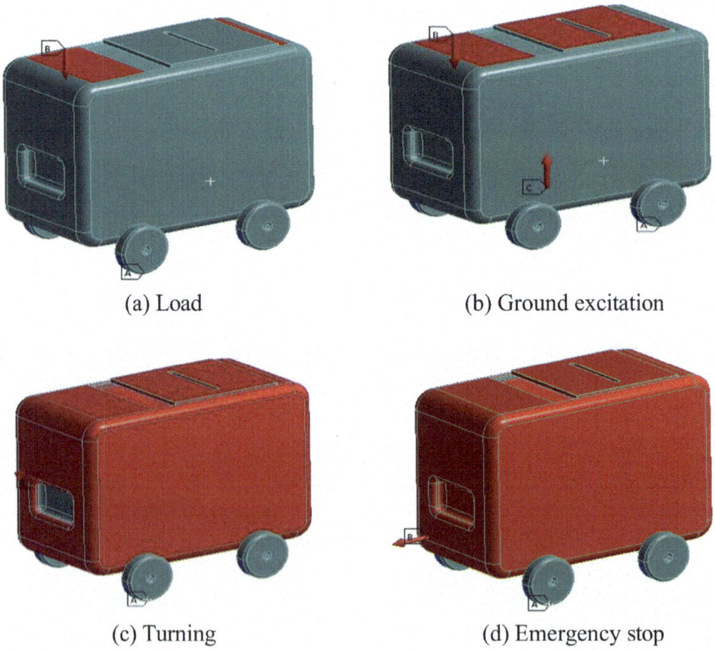

(a) Load (b) Ground excitation

(c) Turning (d) Emergency stop

Fig. 19.4 Establishment of boundary conditions for multiple operating conditions. (**a**) Load. (**b**) Ground excitation. (**c**) Turning. (**d**) Emergency stop

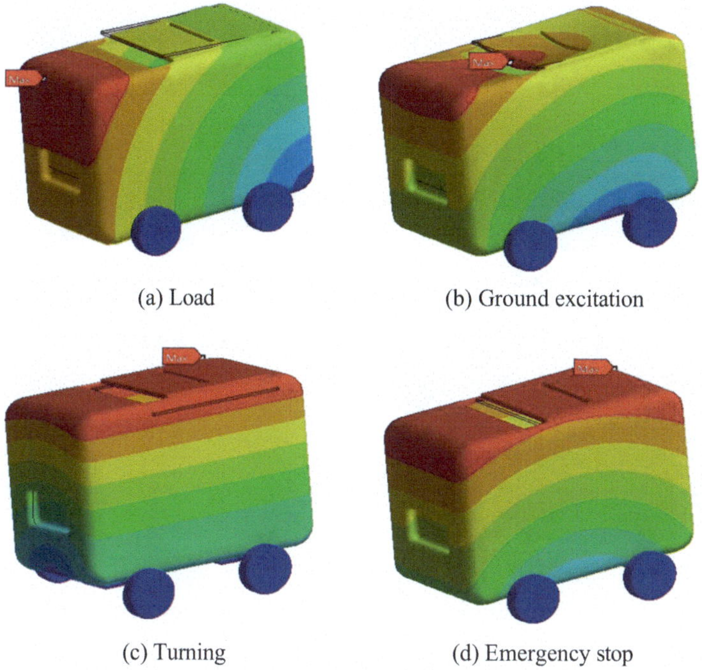

Fig. 19.5 Force clouds under various boundary conditions. (**a**) Load. (**b**) Ground excitation. (**c**) Turning. (**d**) Emergency stop

The maximum total deformation is 1.9 mm under a load of 500 N, 1.258 mm under a ground excitation of 100 N, and 2.412 mm under a turning condition with a force of 100 N. A force analysis was performed for the emergency stop under working conditions, with a force setting of 750 N. The maximum total deformation is 1.389 mm. The strain clouds for the four working conditions are shown in Fig. 19.5.

From the simulation results, it can be seen that the maximum stress is 0.056 Mpa at a pressure of 500 N. The maximum stress is 0.059 Mpa at a force size of 100 N conducted by the ground. The maximum stress is 0.078 Mpa when the transmission trolley is subjected to a bending force of 100 N. The force analysis of the emergency stop was performed, and the maximum stress is 0.035 MPa. The strain cloud can most visually reflect the location of the maximum stress, as shown in Fig. 19.6.

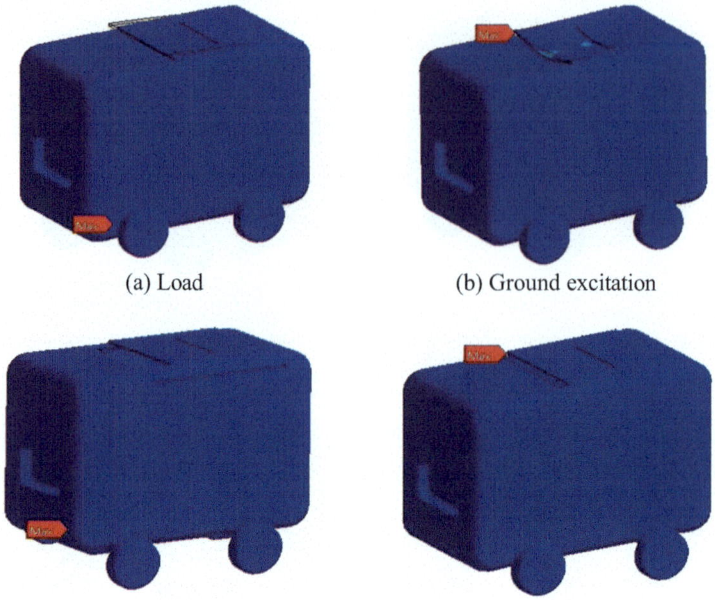

(a) Load (b) Ground excitation

(c) Turning (d) Emergency stop

Fig. 19.6 Strain under various stress conditions. (**a**) Load. (**b**) Ground excitation. (**c**) Turning. (**d**) Emergency stop

19.5 Finite Element Modal Analysis

In the vibration diagram, the intrinsic frequency is usually marked with a number, indicating the vibration period of the structure at that frequency. By observing the intrinsic frequencies in the vibration diagram, the vibration characteristics of the structure can be understood, including the free vibration frequency and resonance frequency. The first ten orders of the intrinsic frequencies are shown in Fig. 19.7. The magnitude of the first-order intrinsic frequency is 2.507 Hz.

Some possible properties for analyzing the maximum deflection of prestressed first- to sixth-order modes are explored. The goal is to move as far away from the system's intrinsic frequency as possible and to avoid resonance. The effect of prestressing is considered in the practical application of handling vehicles. The relevant data were obtained by software simulation, as shown in Table 19.2.

The first six orders of the model pre-stress mode are shown in Fig. 19.8. It can be obtained that the maximum deflection of the short-distance transport trolley mainly occurs at the front end and the bottom rear end of the model from the figure. The speed of the transport trolley is roughly 10–20 km/h. The inertia force is independent of the speed, while the low speed cannot generate large kinetic energy. The place where the maximum deflection occurs is not the main stress point of the trolley, so the design meets the working requirements.

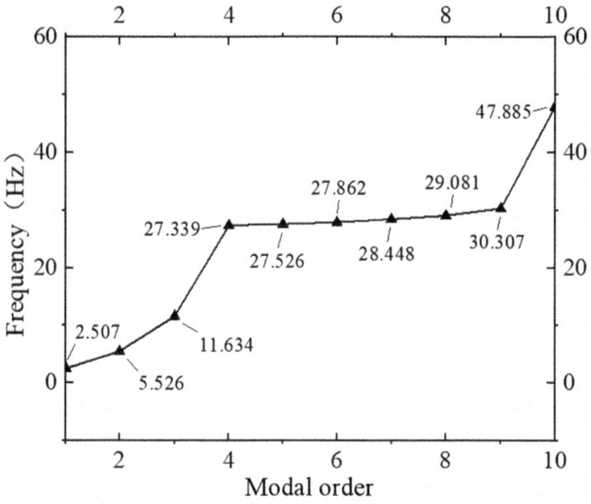

Fig. 19.7 The first 10 steps of the transport vehicle are natural frequency

Table 19.2 Selection of stepper motor

Model	Frequency (Hz)	Maximum deformation (mm)	Vibration characteristics
1	14.3	0.79	Slight deformation at the front
2	17.6	0.73	Tends to deform to the right
3	21.2	0.75	The back end begins to deform
4	25.1	0.74	Chassis arching
5	55.3	0.59	Chassis twisted to the right
6	116.0	0.83	Upper end deformation

According to the above results, the first-order intrinsic frequency of the system is 2.507 Hz, while the overall intrinsic frequency of the system ranges from 11.3 Hz to 116 Hz. This mutual staggering of frequencies ensures that the system does not experience serious resonance phenomena under normal operating conditions. The advantages of this design enable the system to effectively cope with various external disturbances and changes in practical applications, maintaining its normal operation and performance.

19.6 Conclusion

The research results of this paper for the four-wheeled vegetable delivery trolley for short-distance delivery are as follows:

In this paper, we used SolidWorks drawing software to model and assemble the parts of the vegetable transport truck in 3D.

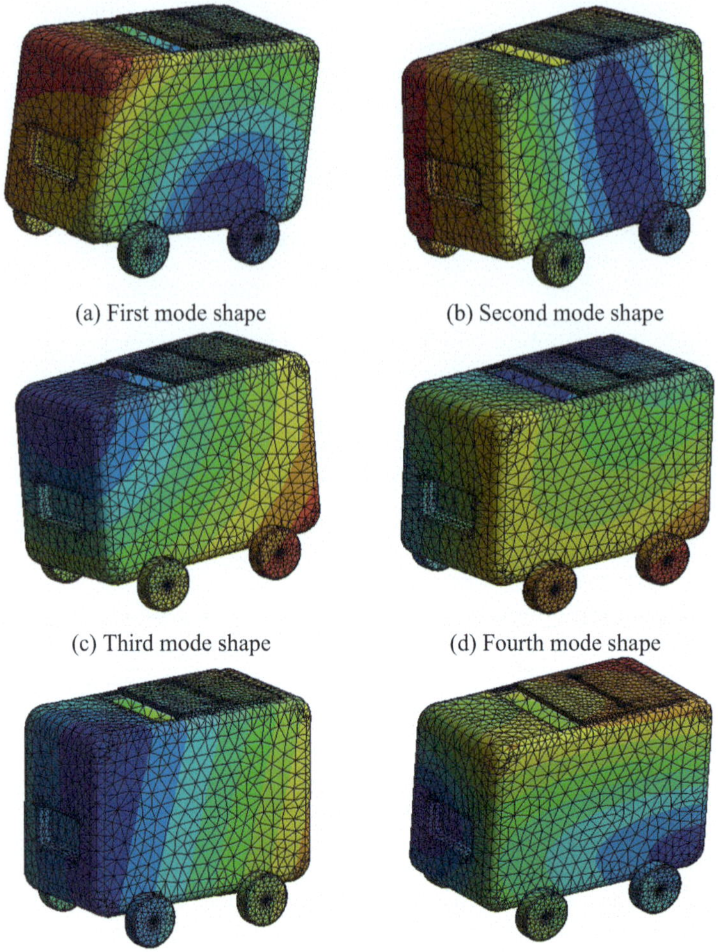

(a) First mode shape
(b) Second mode shape
(c) Third mode shape
(d) Fourth mode shape
(e) Fifth mode shape
(f) Sixth mode shape

Fig. 19.8 First six steps of vibration pattern of the transport vehicle. (**a**) First mode shape. (**b**) Second mode shape. (**c**) Third mode shape. (**d**) Fourth mode shape. (**e**) Fifth mode shape. (**f**) Sixth mode shape

The selected material is 45 steel, and the forces under different working conditions are analyzed after meshing in workbench.

The maximum deformation of the short-distance delivery trolley under different working conditions is 0.078Mpa. The modal analysis of the first six orders of the finite element of the short-distance delivery trolley is carried out, and the deformation is within 1 mm. The allowable stresses are in accordance with the structural modeling of this design. The results show that the designed delivery cart meets the requirements for use under actual working conditions in normal working environment.

Acknowledgments This entire study was funded by the Industrial Technology Development and Application Project of Fujian Province (Project No.2021I0024).

References

1. Chen, Z.B., Zhou, Y.: Brief report on research works of seven papers. Trans. Chin. Soc. Agricult. Mach. **36**(7), 153–157 (2005)
2. Qi, D.Z., Ruan, X.S., Wu, Y.Z., et al.: Optimal design of vibration damping system for transport vehicles between rows in mountain orchards. J. Agricult. Mech. Res. **45**(10), 237–242 (2023)
3. Qi, Y.N., Chen, N., Li, P.: Direct yaw-moment control on four-wheel steering vehicles. J. Southeast Univ. (Nat. Sci. Edit.). **34**(4), 451–454 (2005)
4. Wang, Y.J., Hu, R., He, M., et al.: Design and analysis of a monorail transporter with speed adaptability for mountainous region. Mach. Des. Manuf. **379**(9), 197–201 (2022)
5. Lin, C.: Multi-body dynamics simulation and experimental study on four-wheel-steering vehicle. Automot. Eng. **27**(2), 212–216 (2005)
6. Wang, W.L., Xu, C.C., Wang, Z.H., et al.: A study on finite element analysis and structural assessment method for aluminum-alloy vehicle body. Automot. Eng. **41**(6), 607–614 (2019)
7. Ding, Y.Q., Ma, Y.M.: Finite element of strength and stiffness of car body. J. Army Eng. Univ. PLA. **13**(3), 320–323 (2012)
8. Su, C.Q., Lv, Z.H., Zhang, Q.: Finite element analysis on the lateral crashworthiness of car body structure. Automot. Eng. **29**(11), 964–969 (2007)
9. Gang, X.Y., Zhang, F., Chai, S., et al.: FE modeling method for an air suspension bus under turning and torsion conditions. Automot. Eng. **35**(12), 1099–1104 (2013)
10. Zhao, G., He, Z.G., Zhang, M.Z., et al.: FEM analysis of the abrupt-turning of low-floor rear engine articulated bus. J. Highw. Transp. Res. Dev. **23**(7), 142–146 (2006)

Chapter 20
Research on Parameter Matching and Optimization of Power Transmission System of Pure Electric Logistics Vehicle

Jun-Xiong Liu, Guo-Fei Yu, Run-Fa Chen, and Chao Liang

Abstract AMT pure electric vehicles have the advantages of convenience. Optimizing the pure electric vehicle logistics transmission system can improve the vehicle performance and driving distance. Firstly, according to the dynamic performance, economy and vehicle parameters, the pure electric logistics vehicle's power system parameters are preliminarily matched. Secondly, the model of electric logistics vehicle was established by Cruise simulation platform, and the simulation is carried out according to the dynamic and economic indicators of this paper. Finally, the transmission ratio is used as an optimization variable to establish a multi-objective optimization function and optimize the gear ratio based on the genetic algorithm. The optimized transmission ratio was introduced into the simulation model for simulation. After simulation, it can be seen from the results that the maximum speed is 103.88 km/h to 109.06 km/h, which increases by 4.9%. 0–50 km/h acceleration time is shortened by 1.1% from 8.58 s to 8.483 s. Acceleration time of 50–80 km/h decreased by 4% from 14.68 s to 14.08 s. The maximum stable climbing slope of the first gear increased from 31.14% to 32% and 2.7%; The maximum driving range of 60 km/h constant speed increased from 254.476 km to 260.145 km by 2.2%. The power and economy of the vehicle have been improved.

20.1 Introduction

The use of energy, such as oil, gas, and coal, has an impact on the economy and the progress of science and technology. But there are also energy shortages and competition for resources. Excessive use of energy leads to environmental pollution and other problems. In order to reduce our dependence on oil, it is urgent to develop

J.-X. Liu · G.-F. Yu (✉) · R.-F. Chen · C. Liang
School of Mechanical and Automotive Engineering, Xiamen University of Technology, Xiamen, China
e-mail: gfyu@xmut.edu.cn

S.-P. Tseng et al. (eds.), *Advances in Intelligent Information Hiding and Multimedia Signal Processing, Volume 2*, Smart Innovation, Systems and Technologies 416, https://doi.org/10.1007/978-981-97-8760-9_20

pure electric vehicles. At present, driving range is still an important factor affecting the development of pure EVs. So the research on transmission system parameter matching of pure electric vehicles is particularly important.

At present, in order to solve the problem described above, scholars have carried out in-depth research. Li et al. [1] show that after joining the AMT, the power performance and fuel economy of pure electric vehicles will be improved. Fu et al. [2] present a multi-objective optimization based on hybrid optimization parameter matching method, which can significantly decrease fuel consumption and vehicle emissions, provide the optimal vehicle configuration for the research on energy management strategy. Lin et al. [3] took an electric vehicle as the research object and matched the power system parameters. The results showed that while increasing the number of batteries increased driving range, it also affected the performance. Increasing the power of motor can only increase the power to a limited extent and reduce the economy; proper transmission ratio can improve economy and power performance. In Li et al. [4], pure electric vehicles in cities are used as the research object. Through the vehicle parameters, performance, and fuel economy requirements on vehicle power system parameters matching, the genetic algorithm can optimize the matching of power system parameters, and finally get the optimal parameters of the power system. According to Gao et al. [5], the combination of a pure electric vehicles, such as the requirement of design, matching design of the vehicle power system, multi-objective genetic algorithm is proposed based on a pure electric vehicle parameters optimization scheme, and the optimal velocity ratio is obtained. Liu et al. [6] chose particle swarm optimization algorithm to optimize the parameters of the power and transmission system of the extended range electric vehicle, and the optimized design parameters can significantly improve the technical indicators of the vehicle.

This research studies the pure electric logistics vehicle that was selected as the research object, matches and optimizes the power system parameters, establishes multi-objective optimization function with the main reducer transmission ratio and gear transmission ratio as optimization variable, solves transmission ratio through genetic algorithm, optimal transmission ratio, and then the simulation verifies that the performance and fuel economy of the pure electric logistics vehicle have been improved.

20.2 Performance Metrics and Basic Parameters of EV

The powertrain system is mainly made up of storage battery, gearbox, motor, differential and main reducer. These components can be arranged on the vehicle chassis according to the design requirements. Electromechanical integrated drive arrangement is a commonly used powertrain arrangement scheme, as shown in Fig. 20.1.

According to GB1589–2016 restrictions on vehicle dimensions, axle load and mass, market requirements for EVs, and the regulations of GB/T28382–2012 [7], the study of EVs' performance metrics and basic parameters is shown in Table 20.1.

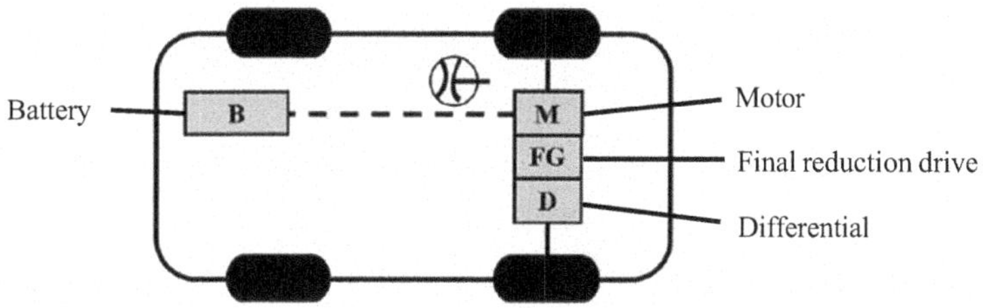

Fig. 20.1 Structure of power drive system of pure electric vehicle

Table 20.1 Basic parameters of pure electric logistics vehicles

Vehicle basic parameters	Numerical value
Length × width × height *mm*	995 × 2200 × 3150
Quality of maintenance *m0/kg*	3010
Full load mass *ma/kg*	4495
Wheelbase *L/m*	3.665
The distance from the center of mass to the front axis *a/m*	1.466
The distance from the center of mass to the rear axis *b/m*	2.199
Height of center of mass *hg/m*	0.8
Windward area A/m^2	3.2
Rotational mass conversion factor δ	1.05
Air drag coefficient *CD*	0.36
Tire rolling radius *r/m*	0.372
Rolling resistance coefficient *f*	0.01
Mechanical efficiency of drivetrain η_T	0.9
Maximum speed umax/km · h^{-1}	≥100
Acceleration time t/s	0–50 km/h acceleration time ≤ 10 50–80 km/h overtaking acceleration time ≤ 15
Maximum climbing grade imax/%	30 (speed≥15 km/h)
Maximum driving range S/km	≥250 (60 km/h even-speed driving)

20.3 Powertrain Parameters Matching

20.3.1 Motor Parameters Matching

The parameters of motor directly influence the vehicle power and fuel economy, so the matching motor parameters can not only consider the economy without considering the power, otherwise all the work will be meaningless. Several important parameters affecting the drive motor performance are: rated power, maximum speed, rated speed, maximum torque, and rated torque. These parameters are matched in turn.

1. The Motor Rated Power

When the electric vehicle runs at the highest speed, motor power should meet the requirements:

$$P_{umax} = \frac{u_{umax}}{3600\eta_T}\left(mgf + \frac{C_D A u^2}{21.15}\right) \tag{20.1}$$

where P_{umax} is the motor rated power, u is vehicle velocity, km/h, g is gravity acceleration, η_T is mechanical transmission efficiency, C_D is wind resistance coefficient, f is rolling resistance coefficient, A is windward area, m^2, α is slope angle, and δ is vehicle rotation mass conversion factor.

When the electric vehicle is climbing, motor power should meet the following requirements:

$$P_{\lambda max} = \frac{1}{\sigma_t}\left(\frac{Gf\,cos\alpha}{3600}u_a + \frac{C_D A u^3}{76140} + \frac{G sin\alpha}{3600}u_a\right) \tag{20.2}$$

The power required to meet the acceleration performance is:

$$P_{ma} = \frac{1}{\sigma_t}\left(\frac{Gf}{3600}u + \frac{C_D A u^3}{76140} + \frac{\delta m}{3600}\frac{du}{dt}\right) \tag{20.3}$$

To sum up, the motor rating of an electric vehicle is:

$$P_{max} = \max\{P_{umax}, P_{\lambda max}, P_{ma}\} \tag{20.4}$$

2. Motor Speed

The maximum working rotational speed of motor shall meet the basic speed of motor:

$$\beta = \frac{n_{max}}{n_e} \tag{20.5}$$

where n_e is the motor base speed (rated speed), n_{max} is the motor highest speed, and β is the constant power region coefficient of motor. β is set at 2.31, so the design of the motor rated speed is 2810 r/min.

3. Measurement of the Motor Torque.

Power and speed can be calculated by rated torque of the motor and peak torque of the motor, and the solution expression is:

$$T_e = \frac{9550P_e}{n_e} \tag{20.6}$$

$$T_{max} = \lambda \cdot T_e \tag{20.7}$$

By substituting the rated speed and rated power into Eq. (20.6), motor rated torque T_e can be calculated as 136 $N \cdot m$, then motor peak torque T_{max} can be obtained as 272 $N \cdot m$ from Eq. (20.7).

20.3.2 Transmission System Parameter Matching Design

The minimum transmission ratio range is expressed as follows:

$$\frac{rn_{max}}{9500P_{max}\eta_T}\left(mgf + \frac{C_DAu^2}{21.15}\right) \le i_{tmin} \le 0.377\frac{rn_{max}}{u_{max}} \tag{20.8}$$

The maximum transmission ratio range is expressed as follows:

$$i_{tmin} \le \frac{Mgr\varphi}{T_{max}\eta_T}\left(\frac{b}{L}\cos\alpha_{max} - \frac{h_g}{L}\sin\alpha_{max}\right) \tag{20.9}$$

where h_g is the height of vehicle mass center; L is automobile wheelbase; b is the distance between rear wheel and mass center; φ is the average adhesion coefficient of pavement; T_{max} is the motor maximum torque, $N \cdot m$; α_{max} is the maximum climbing angle; η_T is drivetrain efficiency.

The gear of transmission system directly influences the power and fuel economy of EVs. According to the method of selecting and developing shifting rules, combined with the actual needs of enterprises, considering that the research object of this paper is the pure electric logistics vehicle driving in mountain road and basin road conditions, the range and climbing performance of the vehicle should be taken into account [8]. This paper determines the gear of transmission as four gears.

20.3.3 Power Battery Parameter Matching

Assuming that the electric logistics vehicle travels at an even velocity of $v_a = 60$ km/h, the required power P_m can be calculated by the following formula:

$$P_m = \frac{1}{\eta_T}\left(\frac{mgfv_a}{3600} + \frac{C_DAv_a{}^3}{76140}\right) \tag{20.10}$$

The result is $P_m = 10.44$ kW.

According to the economic indicators of EVs, the distance S of EVs running at a constant speed of $v_a = 60$ km/h after a single charge shall not be less than 250 km:

$$Q_t = \frac{P_m S}{v_a \eta_b \eta_m} \tag{20.11}$$

where S is 250 km, Q_t is the electric energy consumed by the vehicle during running, kW· h, η_b is efficiency of battery pack is 0.9, and η_m is the efficiency of the driving motor is 0.95. It is calculated that $Q_t = 50.88$ kW · h.

And because the battery can provide at least more energy than the car in the process of consumption can be obtained:

$$E = \frac{UC}{1000} \geq Q_t \tag{20.12}$$

where U is the total voltage of battery, V; E is battery pack energy, kW · h, C is battery capacity (Ah). According to calculation, C is greater than 88.34 Ah at least. Considering the loss of electric power to electrical components, the actual battery capacity C is 100 Ah.

20.4 Modeling and Analysis of Vehicle

After parameter matching is completed, the rationality of parameter matching results should be verified. In this chapter, the simulation model of pure electric logistics vehicle will be established by AVL Cruise software to simulate and analyze the vehicle power and economy under the four-speed AMT scheme [9].

This paper chooses to modify the existing model to get the desired simulation model. In addition to the original components, it is also necessary to establish the corresponding electronic control system of each module, such as: shift control system, drive control system, brake control system, etc. In this paper, except for the shift control strategy formulated by oneself, other control systems adopt software default settings.

Figure 20.2 shows the simulation results of the maximum speed. As can be seen from Fig. 20.2, maximum speed can reach 103.88 km/h, meeting the dynamic performance index of the maximum speed not less than 100 km/h.

Four-speed AMT electric car logistics 0–50 km/h and 50–80 km/h speed number as shown in Figs. 20.3 and 20.4, respectively. As can be seen from Fig. 20.5, 0–50 km/h acceleration time is 8.58 s. Figure 20.6 shows that 50–80 km/h overtaking acceleration time is 14.68 s. Both meet the design requirements of acceleration performance.

The first-gear climbing slope curve of the vehicle under full load is shown in Fig. 20.5. When the vehicle is climbing at 15 km/h, the maximum climbable gradient can reach 30.891%, which meets the climbing requirements designed in this paper.

Figure 20.6 reflects the relationship between driving range and SOC of pure electric logistics vehicle with time under 60 km/h cycle condition. It is known that

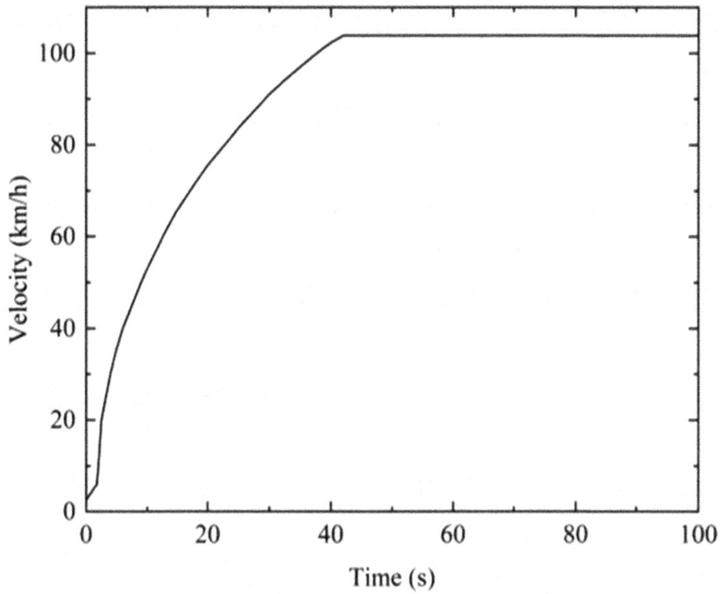

Fig. 20.2 Maximum speed of four-gear AMT

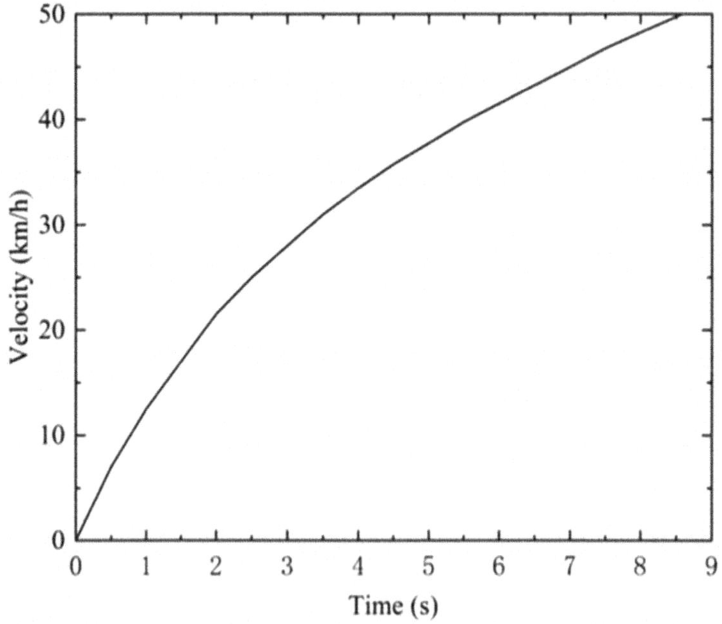

Fig. 20.3 Acceleration time of 0–20 km/h for four-gear AMT

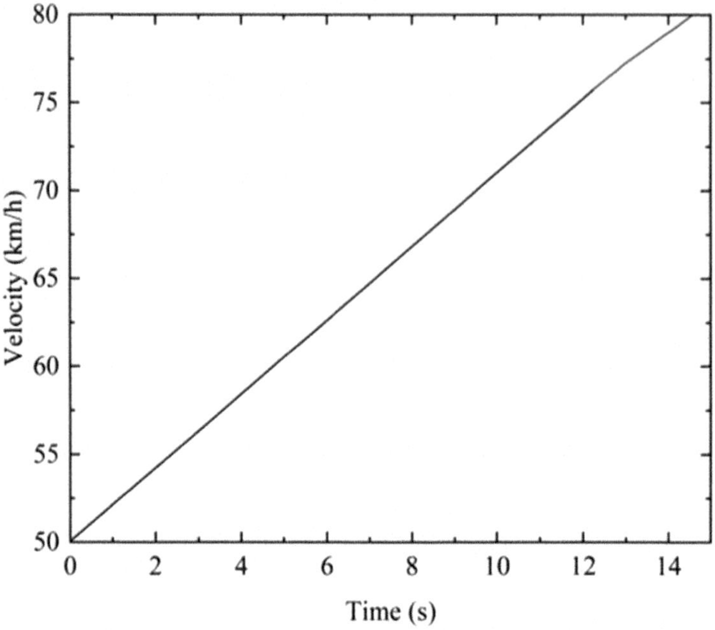

Fig. 20.4 Acceleration time of 50–80 km/h for four-gear AMT

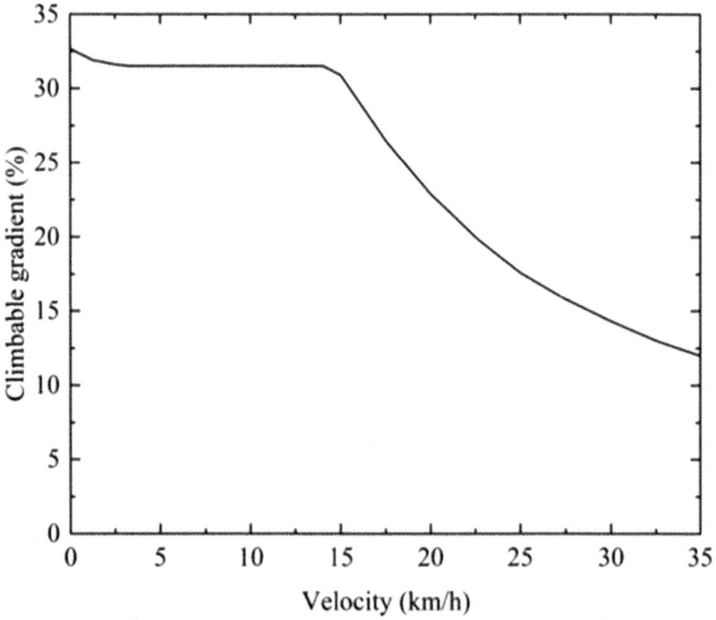

Fig. 20.5 Maximum climbable gradient of the first gear of the fourth gear AMT

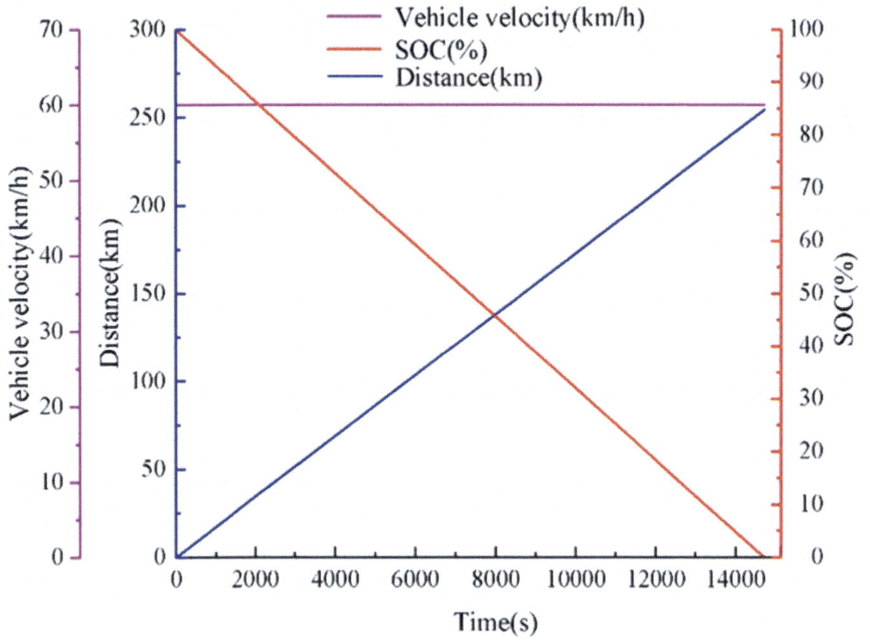

Fig. 20.6 Change of driving range and electric quantity of four-gear AMT at 60 km/h constant speed

the maximum driving distance under this condition can reach 254.476 km, meeting the economic design index.

20.5 Transmission System Parameter Optimization

20.5.1 Establishment of Objective Function

Maximum speed and driving distance were taken as objective functions. Before transforming a multi-objective problem into a single objective problem, it is necessary to unify the dimensions, because without unification, two different objective functions cannot be evaluated by adding. After quantization, the multi-objective optimization function is:

$$f(x) = w_1 \frac{f_1(x)}{v_e} + w_2 \frac{f_2(x)}{S_e} \tag{20.13}$$

where v_e is the maximum speed before optimization, km/h; S_e is 60 km/h constant speed mileage before optimization; w_1 and w_2 are the weight coefficients, both are positive numbers, and $w_1 + w_2 = 1$. According to different design requirements, the weight coefficient of w_1 is temporarily set at 0.4 and w_2 at 0.6.

20.5.2 Establishment of Constraints

According to the request of the power and economy of vehicle and transmission ratio distribution rule, the following constraints are established:

1. Maximum speed shall not be lower than 100 km/h, and the motor power under the maximum speed shall not exceed the maximum power of the motor, then:

$$g_1(x) = 0.377\frac{rn_{max}}{i_0 i_{g4}} - 100 \geq 0 \tag{20.14}$$

$$g_2(x) = mgf\,v_{max} + \frac{C_D A u^2}{21.15} - P_{max} \geq 0 \tag{20.15}$$

2. If the vehicle is running at the speed of 15 km/h and its maximum climbing slope is not less than 30%, then:\vspace*{-10pt}

$$g_3(x) = \tan\left[\arcsin\left(\frac{F_t - F_w}{Mg\sqrt{f^2+1}}\right) - \arcsin(f)\right] - 0.3 \geq 0 \tag{20.16}$$

3. The maximum mileage of 60 km/h constant speed shall not be less than 250 km, then

$$g_4(x) = \frac{E v_a i_0 i_{g1} \eta_T}{P_m} - 250 \geq 0 \tag{20.17}$$

4. In order to prevent skidding when driving, the pure electric logistics vehicle should meet the adhesion conditions, then:

$$g_5(x) = \varphi F_z - \frac{T_{max} v_a \eta_b \eta_m}{t} \geq 0 \tag{20.18}$$

5. The transmission ratio between each adjacent gear of the gearbox is usually distributed according to the geometric series. To avoid the transmission ratio

between adjacent gears being too large, leading to the difficulty of shifting, there are:

$$g_6(x) = 1.4i_{g2} - i_{g1} \geq 0 \tag{20.19}$$

$$g_7(x) = i_{g1} - 1.75i_{g2} \geq 0 \tag{20.20}$$

$$g_8(x) = 1.4i_{g2} - i_{g2} \geq 0 \tag{20.21}$$

$$g_9(x) = i_{g2} - 1.75i_{g3} \geq 0 \tag{20.22}$$

$$g_{10}(x) = 1.4i_{g4} - i_{g3} \geq 0 \tag{20.23}$$

$$g_{11}(x) = i_{g3} - 1.75i_{g4} \geq 0 \tag{20.24}$$

20.5.3 Optimization Result

The transmission ratio fitness function is compiled in MATLAB, and the transmission ratio is optimized by genetic algorithm. The final transmission ratio optimization results based on genetic algorithm are shown in Table 20.2.

20.5.4 Simulation Analysis and Comparison After Optimization

The optimized transmission ratio parameters are substituted into AVL Cruise vehicle model, and the dynamic and economic simulation calculations are carried out. The results obtained will be compared with the data before optimization to evaluate the optimization effect [10].

Table 20.2 Optimized transmission ratio parameters of the front and rear drivetrains

Parameter	After optimization	Before optimization
Transmission I gear ratio i_{g1}	4.52	4.5
Transmission II gear ratio $i_{g\,2}$	2.58	2.54
Transmission III gear ratio $i_{g\,3}$	1.62	1.63
Transmission IV gear ratio $i_{g\,4}$	0.97	1
Main reducer transmission ratio i_0	5.26	5.21

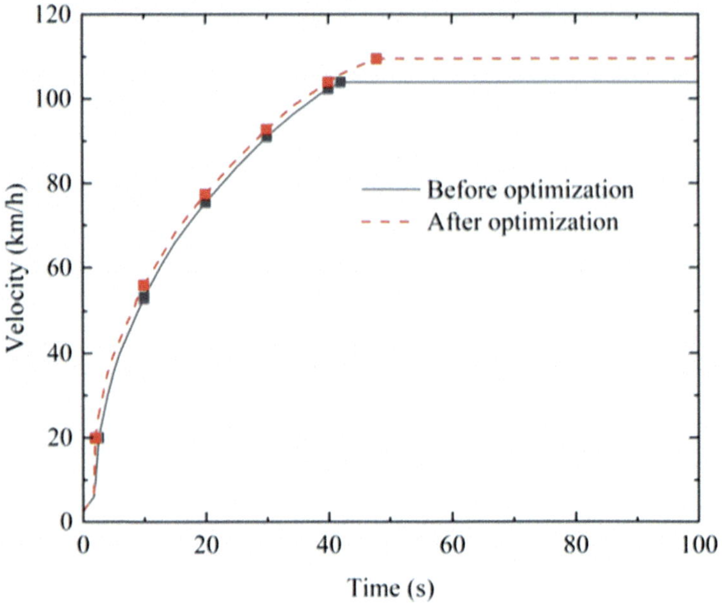

Fig. 20.7 Comparison of maximum speed before and after optimization

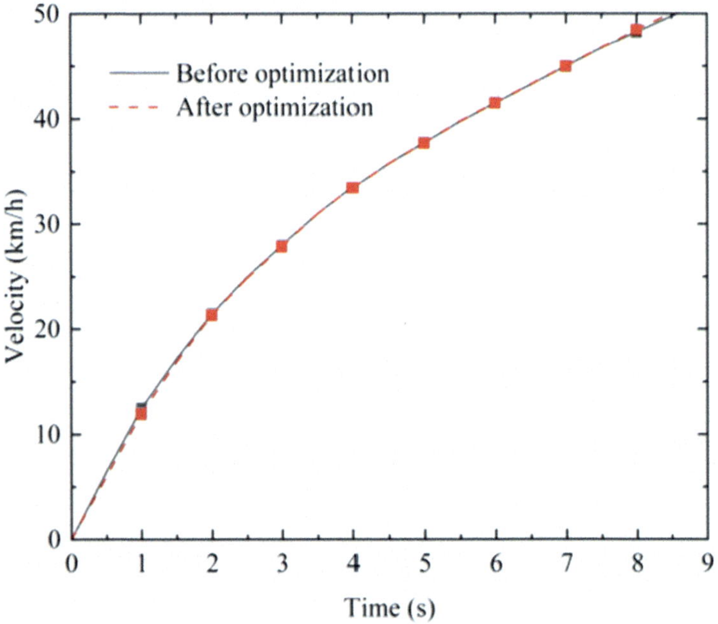

Fig. 20.8 Comparison of 0–50 km/h acceleration time before and after optimization

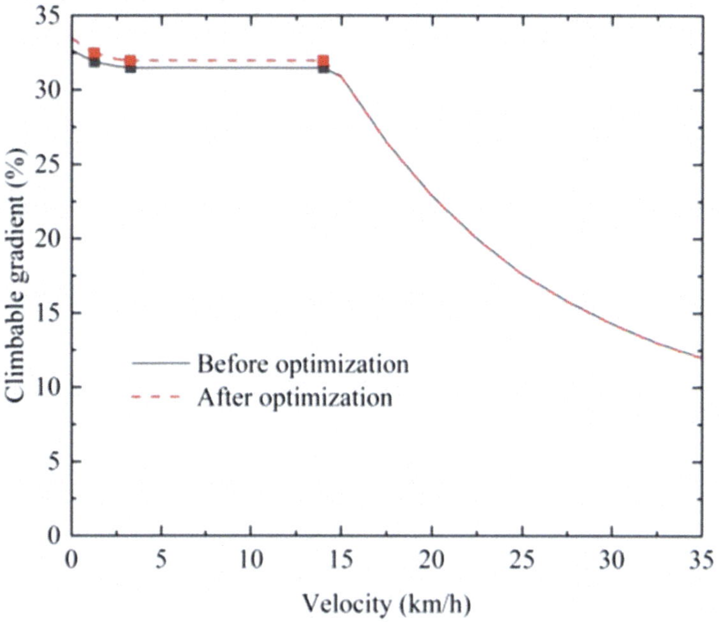

Fig. 20.9 Comparison of 50–80 km/h acceleration time before and after optimization

Figure 20.7 shows the comparison of the maximum speed before and after the optimization of ratio of transmission parameters of transmission system of four-speed vehicle. After optimization, the maximum speed increases from 103.88 km/h to 109.06 km/h. The comparison of 0–50 km/h acceleration time is shown in Fig. 20.8. The acceleration time decreases from 8.58 s before optimization to 8.483 s. The comparison of acceleration time of 50–80 km/h is shown in Fig. 20.9, and the acceleration time is reduced from 14.68 s before optimization to 14.08 s. Comparison of maximum climbable gradient is shown in Fig. 20.10. After optimization, the maximum climbable gradient of the first gear increases from 32.67% to 32.86%, and the maximum climbable gradient increases from 31.14% to 32%. When velocity is 15 km/h, the optimized uphill slope of 30.876% still meets the requirements of climbing slope of not less than 30%.

For the convenience of analysis, the simulation results and test data before and after optimization are collated in Table 20.3. Table 20.3 shows that the power and economy of the optimized vehicle are improved. This also verifies the feasibility of using genetic algorithm to optimize transmission ratio parameters of drivetrain in this paper. Compared with the test results, although there are some errors, the errors are relatively small and within the acceptable range.

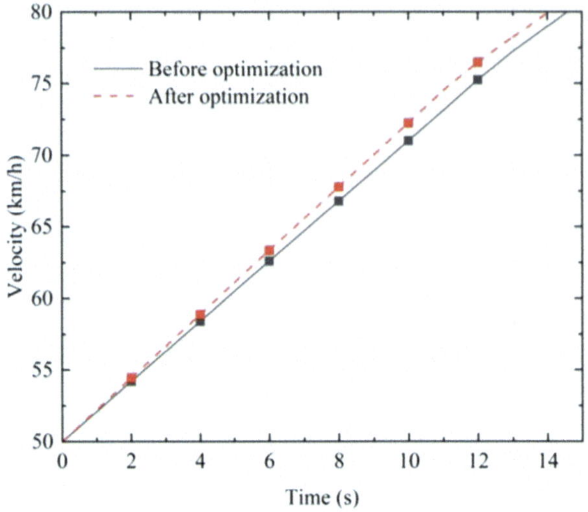

Fig. 20.10 Comparison of maximum climbable gradient before and after optimization

Table 20.3 Comparison of test data before and after optimization

Evaluation method	Before optimization	After optimization
Evaluation index		
Maximum speed (km/h)	103.88	109.06
0–50 km/h acceleration time (s)	8.58	8.483
50–80 km/h acceleration time (s)	14.68	14.08
Maximum climbing grade (%)	32.67	32.86
60 km/h constant mileage (km)	254.476	260.145

20.6 Conclusion

The research object is a vehicle in this paper, parameters matching and optimization of its dynamical system. The results are as follows:

1. According to the power and economic design requirements, the driving motor, transmission ratio, power battery, and other parameters are preliminarily matched, and the motor and battery are selected the four-gear AMT was selected and the transmission ratio was allocated.
2. Build a four-speed AMT pure electric logistics vehicle model with Cruise software, input relevant parameters of the model into each module, and then conduct dynamic and economic simulation analysis. After simulation results, it can be shown that maximum speed can attain 103.88 km/h, The acceleration time of 0–50 km/h is 8.58 s. The overtaking acceleration time of 50–80 km/h is 14.68 s, the maximum gradient of the first gear can reach 32.67%; the maximum

driving range can reach 254.476 km at 60 km/h constant speed, and the vehicle performance and fuel economy meet the performance index requirements.

3. Using genetic algorithm to a four-speed AMT transmission ratio of vehicle optimization, and introduces the optimized ratio in the model for power and fuel economy simulation analysis. After the simulation results, it can be concluded that the power and economy of the optimized vehicle are improved.

Acknowledgments This project is sustained by the Fujian Province Technology innovation key research and industrialization project: Integrated intelligent wire control distributed drive and vehicle package integrated vehicle technology and industrialization (No.: 2022G043), 2022.05-2023.04 (Collaborating unit).

References

1. Li, B.: Research on Design and Performance of AMT Shifting System Based on Direct Drive Technology. Nanjing University of Science and Technology, Jiangsu (2014)
2. Fu, X.-L., Zhang, Q., Tang, J.-Y., et al.: Parameter matching optimization of a powertrain system of hybrid electric vehicles based on multi-objective optimization. Electronics. **8**(8), 1–18 (2019)
3. Lin, X.-Y., Zhang, S.-B., Feng, Q.-G.: Parameter matching optimization and performance inluencing factors analysis of pure electric vehicle. Mach. Des. Manuf. **4**, 126–135 (2017)
4. Li, W.-W., Sun, H.-H., Xu, X.-W.: Parameter matching and optimization of city pure electric vehicle powertrain system. J. Mech. Trans. **41**(2), 22–26 (2017)
5. Gao, E., Wang, H.-L.: Research on parameter matching and optimization of power system of a pure electric vehicle. Mach. Des. Manuf. **8**, 31–35 (2017)
6. Liu, H.-W., Lei, Y.-L., Fu, Y.: Parameter matching and optimization for power system of range-extended electric vehicle based on requirements. Proc. Inst. Mech. Eng. Part D J. Automob. Eng. **234**(14), 3316–3328 (2020)
7. National Automobile Standardization Technical Committee.: GB/T 28382–2012 Technical Conditions for pure Electric Passenger Vehicles. Standards Press of China, Beijing (2012)
8. Jia, Q.-X., Zhang, H.-X., Zhang, Y.-J.: Parameter matching and performance analysis of a master-slave electro-hydraulic hybrid electric vehicle. PRO. **10**(1664), 1–21 (2022)
9. Chen, Y.-N., Xie, B., Du, Y.-F.: Powertrain parameter matching and optimal design of dualmotor driven electric tractor. Int. J. Agric. Biol. Eng. **12**(1), 33–41 (2019)
10. Wu, P., Qiang, P.-H., Pan, T.: Multi-objective optimization of gear ratios of a seamless three-speed automated manual transmission for electric vehicles considering shift performance. Energies. **15**(11), 4149 (2022)

Chapter 21
Research on Classification and Recognition of Atayal Costume Patterns Based on Deep Learning

Jing-Wan Huang, Ke-Ke Sun, and Yu-Yang Yuan

Abstract As one of the many indigenous minority groups with a long history in Taiwan, the simple styles and rich patterns of the Atayal costumes vividly reflect the historical and cultural heritage of their ethnic origin. Based on this cultural background, this paper takes the traditional costume patterns of the Atayal as the research object, uses deep learning, applies the YoloV3 algorithm model for pattern intelligent recognition, compares it with the EfficientDetD0 model, finally analyzes the experimental results by analogy, and concludes that the YoloV3 model used in this paper has a better recognition effect. The results are compared with the EfficientDetD0 model, and the YoloV3 model used in this paper has a better recognition effect.

21.1 Introduction

Researchers in the field of apparel have proposed various methods for using deep learning to train and classify clothing data with different models, which evaluate the advantages and disadvantages of different clothing feature extraction models based on image and design attributes. Guan et al. [1] proposed three methods to train and classify clothing data with different models, which can evaluate the advantages and disadvantages of different clothing feature extraction models through image and design attributes. Another example is the vector-based user preference fashion recommendation model developed by Jo [2], which utilizes implicit analysis for efficient fashion product search and recommendation. Another example is the fashion style matching framework proposed by Liu [3], which models a flexible

J.-W. Huang · K.-K. Sun (✉)
School of Design Art, Xiamen University of Technology, Xiamen, China
e-mail: sunkeke@xmut.edu.cn

Y.-Y. Yuan
School of Mechanical and Automotive Engineering, Xiamen University of Technology, Xiamen, China

S.-P. Tseng et al. (eds.), *Advances in Intelligent Information Hiding and Multimedia Signal Processing, Volume 2*, Smart Innovation, Systems and Technologies 416, https://doi.org/10.1007/978-981-97-8760-9_21

set of fashion items through a graphical neural network and improves the AUC criteria by more than 10%. Quan et al. [4] proposed a deep learning framework that combines product innovation and can transform the information of clothing product style into product shape in real time, establishing a mapping model between product attributes and style. Vijayaraj et al. [5] trained the CNN model architecture on the Fashion-MNIST dataset and adjusted the parameters. Zhang et al. [6] applied CNN to Miao embroidery category recognition, and compared it with three traditional non-depth methods, proving the superiority of CNN model. In this research field, many researchers have taken the comparison or upgrade of algorithm models as an innovation point, and in the field of analog upgrade of algorithm models. Chen [7] developed a method for joint learning of shape descriptors and their corresponding relationships in deep convolutional neural networks, which automatically obtains more discriminative shape descriptors when retrieving clothing styles. In addition, Lin [8] introduced an all-optical diffraction deep neural network architecture approach to achieve classification of handwritten digits and images of fashion products, demonstrating a physical mechanism for performing machine learning. Overall, deep learning is increasingly common in the field of clothing and fashion, and the optimization of algorithms and analogical work is a research hotspot. To expand the research in this area, in this paper, we use deep learning methods to recognize the patterns of the Atayal costume using YoloV3 and compare the recognition effect with that of the EfficientDetD0 model in order to identify a more suitable model for the recognition of Atayal costume patterns.

21.2 Overview and Classification of Atayal Dress Patterns

As one of the three major aboriginal groups in Taiwan, the Atayal have a wide variety of weaving or costume patterns, making their costumes highly researchable. For the Atayal, weaving is not a separate art form, but is inextricably linked to their traditional natural environment, the accumulation of labor and social worship. Pestles, mortars, and stripe, representing rice pounding or beating implements, often appear in weaving motifs, and through the repetitive combination of these motifs, the distinctive patterns of Atayal weaving and dress are expressed. In this paper, the patterns of the Atayal are divided into three broad categories, Shell pearl, Stripe, and Rhombus, based on the shape of the Atayal patterns. In the early collection and labeling process of this paper, these three species were named BZ, TW, and LX respectively.

The Shell pearl pattern is shown in Fig. 21.1. The Shell pearl is made from shells that are ground into small round pieces with an outer diameter of about 3–4 mm, an inner hole of about 1 mm, and a thickness of 1–2 mm, all done by hand from taking the material, grinding the outer circle to drilling the holes. The Shell pearl is stitched onto the ramie fabric in the form of Shell pearl, which is a different pattern from other patterns and is distributed in three dimensions on the garment.

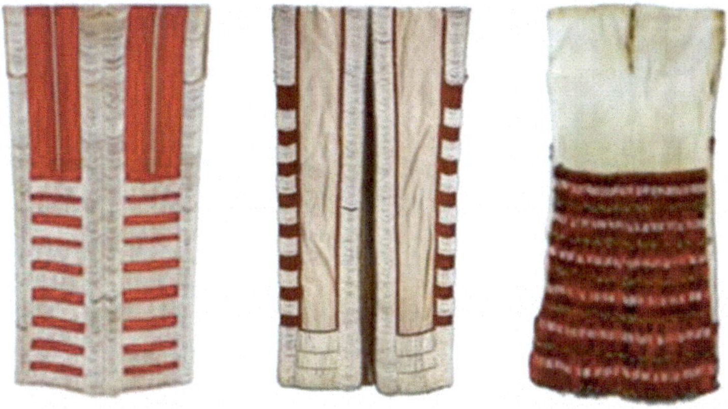

Fig. 21.1 Pattern classification—BZ

Fig. 21.2 Pattern classification—TW

The stripe pattern is shown in Fig. 21.2. Stripe in traditional Atayal culture can represent the implements of their people's daily labor. Unlike stripe made up of simple straight lines, striped patterns in Atayal dress are made up of more detailed individual small patterns circulated and combined into different strips, distributed throughout the dress or woven fabric.

The rhombus pattern is shown in Fig. 21.3. The rhombus, which is a symbol of ancestral eyes and protection for the Atayal people, is also a distinctive feature of Atayal dress patterns, with its various or spreading or circular patterns.

Fig. 21.3 Pattern classification –LX

21.3 Convolutional Neural Networks

21.3.1 Overview of Convolutional Neural Networks

Convolutional neural networks (CNNs) are composed of four main layers: a convolutional layer, a pooling layer, an activation function, and a fully connected layer.

The convolutional layer [7] is considered the key component of the CNN and is responsible for extracting image features. In contrast to a fully connected neural network, CNNs have significantly fewer parameters. The convolutional layer implements the local connection and weight-sharing features of CNNs. In this layer, the convolution kernel comprises the weights and biases. The formula for calculating the output value of the convolution layer is as follows:

$$a_{i,j} = f\left(\sum\nolimits_{m=0}^{2}\sum\nolimits_{n=0}^{2}\omega_{m,n}x_{i+m,j+n} + \omega_b\right) \tag{21.1}$$

In this equation, $x_{i,j}$ represents an element in row i and column j of the input feature image, $\omega m,n$ represents a weight in row m and column n of the convolutional kernel, ωb denotes the bias term of the kernel, ai,j denotes an element in row i and column j of the output feature image, and f denotes the activation function applied to the sum of the weighted inputs and the bias term. This equation represents the fundamental operation of convolutional neural networks, which is used to extract features from input images and classify them into different categories. In fact, the width and height of the resulting feature graph satisfy the relationship as follows:

$$\begin{cases} W_2 = (W_1 - F + 2P)/S + 1 \\ H_2 = (H_1 - F + 2P)/S + 1 \end{cases} \tag{21.2}$$

where W_2 and W_1 represent the width of the feature image after and before convolution, respectively; F denotes the width of the convolution kernel; P represents the number of zero padding; S represents the step size; and H_2 and H_1 represent the height of the feature image after and before convolution, respectively.

Pooling layers can reduce data dimensions, help reduce overfitting during training, and effectively reduce the number of model parameters used for data compression. Average pooling layer and maximum pooling layer are generally used. Due to the slow convergence of convolutional neural networks and the complex computation process, in order to maximize the performance of the network, this paper chooses the maximum pooling layer with relatively simple computation. The calculation formula of maximum pooling is as follows:

$$f(x) = \max\left(x_{[i,i+k][j,j+k]}\right) \tag{21.3}$$

Common activation functions include sigmoid, ReLU, Leaky ReLU, and so on. The commonly used sigmoid function expression is as follows:

$$f(x) = \frac{1}{1 + e^{-x}} \tag{21.4}$$

The ReLU function is a more commonly used activation function in deep learning because it is faster than the sigmoid function and does not suffer from gradient saturation when the input is positive, its expression is as follows:

$$f(x) = \begin{cases} \max(0, x), x \geq 0 \\ 0, x < 0 \end{cases} \tag{21.5}$$

$$P(yx) = \frac{e^{h(x, y_i)}}{\sum_{j=1}^{n} e^{h(x, y_i)}} \tag{21.6}$$

21.3.2 YoloV3 Target Detection Model

YoloV3 [8] is an object detection algorithm, which inputs the entire image into the convolutional neural network (CNN) as a whole and divides the image into multiple grids, each of which is responsible for detecting targets within the grid. Each grid can predict multiple boundary boxes, and each boundary box contains the category of the target, the confidence level, and the coordinates of the boundary box. The confidence calculation formula is as follows:

$$confidence = \Pr(Object) \times IOU_{pred}^{predtnuh} \tag{21.7}$$

where Pr(Object) is the probability of containing a certain type of object in the grid, if it contains a value of 1 otherwise it is 0. IOU indicates the precision of the output bounding box, defined as the intersection ratio of the predicted bounding box to the real bounding box. The pattern object detection process of the algorithm in this paper is shown in Fig. 21.4. The pattern image data is first input, then a deep network is constructed for pattern feature extraction, followed by combinatorial learning of different levels of features and pattern feature mapping models, then feature processing pattern matching localization and logistic regression operations, and finally the test results are output.

21.4 Image Processing and Experimentation

21.4.1 Raw Image Material Collection

This experiment uses Python 3.8.0, the deep learning framework pytorch + OpenCV4 programming to write code to build the model. The pattern image dataset

Fig. 21.4 Atayal pattern target detection process

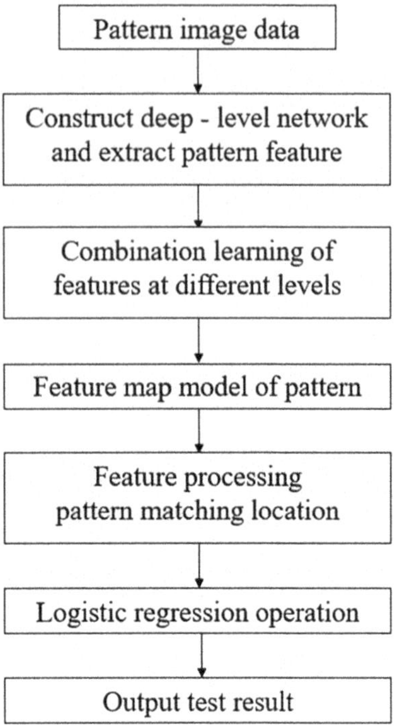

used in the experiment was obtained from three parts, the first part was obtained from field survey shooting, the shooting materials came from the Shunyi Aboriginal Museum in Taipei, the Taiwan Aboriginal Peoples Cultural Park in Pingtung County, and the Tribal Bay Recreation Area in Hualien County; the second part was to crawl the online images of the three types of images, through the Python crawler in Baidu images. The third part is the images obtained by downloading from official websites such as China News, the National Palace Museum and the National Cultural Memory Bank. As the crawled images needed to be read for data and tags, the images were renamed in bulk, and some of the sample images in the Atayal costume pattern dataset are shown in Fig. 21.5.

21.4.2 Data Set Annotation

Prior to image recognition, the Atayal pattern needs to be annotated with the location of the dress or weave to obtain the groundtruth box. Existing target detection models require the use of pre-annotated datasets, so Labelimg was chosen to annotate the targets in the traditional Atayal patterns. The environment python+xml needs to be configured before using Labelimg. Before the annotation the classification of the Atayal pattern for this experiment is preset, as the environment path must be in

Fig. 21.5 Compilation of some original photographic material

Fig. 21.6 Labeling tool demonstration

English, the classification "Stripe", "Rhombus", and the three classification presets for "Shell pearl" are named "TW", "LX" and "BZ". The labeling tool is shown in Fig. 21.6.

21.4.3 Experimental Design Process

The experimental process of image classification is shown in Fig. 21.7. The experimental part of this paper is as follows, batch reading of pattern images, obtaining data and label values; then pre-processing and data enhancement of the

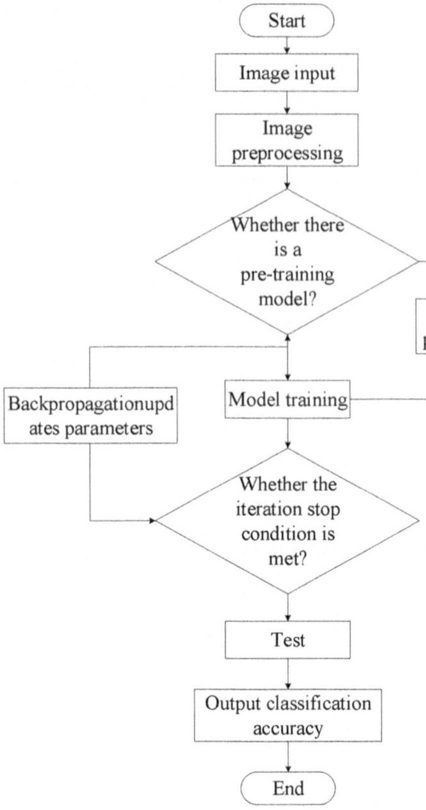

Fig. 21.7 Flow chart of image classification experiment

pattern images, as the original material images of this paper are small, in order to solve this problem, the size of the images and the data enhancement operation of the sample library are needed, retaining the original content of the images, rotating, color changing, and blurring. Figure 21.8 shows the comparison of some of the images before and after the data enhancement operation.

In the classification experiment, the classified images were divided into two parts, test set and training set, where the ratio of test set to training set was 9,1. The order of pattern image data and labels were randomly disordered, and then the training set and test set were divided; after the division, the hyperparameters were set, and before the model experiment was conducted, all hyperparameters were first set uniformly, and the basic parameters of the network were set as shown in Table 21.1.

After setting the hyperparameters the pre-trained model parameters were loaded to build the YoloV3 and EfficientDetD0 models. When model parameters are randomly initialized for model training this increases the time complexity and makes training inefficient. In this case, the pre-training model was created. Instead of train-

(a) Before data enhancement

(b) After data enhancement

Fig. 21.8 Partial image data enhancement process. (**a**) Before data enhancement. (**b**) After data enhancement

Table 21.1 Parameter setting table

Hyperparameters	Data
Freeze_Batch_size	8
Freeze_epoch	0–50
Freeze_learning_rate	1×10^{-3}
Unfreeze_Batch_size	4
Unfreeze_epoch	51–200
Unfreeze_learning_rate	1×10^{-4}

ing the model from scratch, a well-trained model with good classification results can be used as the initial parameters for training the model, a process known as model pre-training. In general, the more images there are, the better the classification results. In this paper, there are only 1059 images of traditional Atayal costumes, and a smaller number of images may be overfitted in the subsequent training process. Therefore, this paper adopts a pre-training strategy that can substantially compress the model training time and improve the classification efficiency. Model training time improved classification efficiency, defined loss, optimizer, and precision for model training. This is followed by determining whether the number of iterations is satisfied and backpropagating the updated parameters. Finally, the model is tested and the model precision is output.

21.5 Analysis of Experimental Results

21.5.1 Comparison of Recognition Effects

The recognition results of the two models for the same dataset under the same conditions of operation are shown in Fig. 21.9. A selection from each of the three categories of the dataset is compared and it can be seen that YoloV3 is significantly better than the EfficientDetD0 model for all three categories, BZ, TW, and LX.

21.5.2 Model Evaluation Criteria

Image classification is evaluated using percision, and the level of percision is a direct indicator of how good the classification results are. In this paper, the main evaluation criteria for image classification are Iteration, AP, Precision, Recall, and F1 value.

Figure 21.10 shows the AP-Recall curves of the two models for the three category identifications. The left side represents the EfficientDetD0 model, and the right side represents the YoloV3 model, ap represents the average accuracy. The horizontal axis represents the recall rate and the vertical axis represents the average accuracy using the logarithmic axis. The overall trend is that the Precision value tends to decrease as the Recall value increases. In Fig. 21.10, the curves of (d), (e), and (f)

(a) EfficientDetD0 Model recognition effect

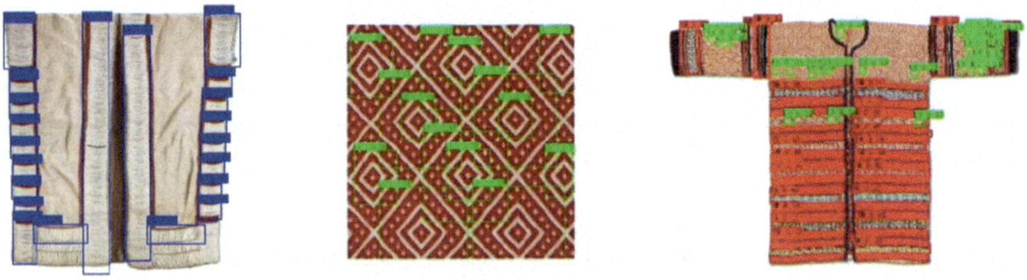

(b) YoloV3 Model recognition effect

Fig. 21.9 Recognition of the two models. (**a**) EfficientDetD0 Model recognition effect. (**b**) YoloV3 Model recognition effect

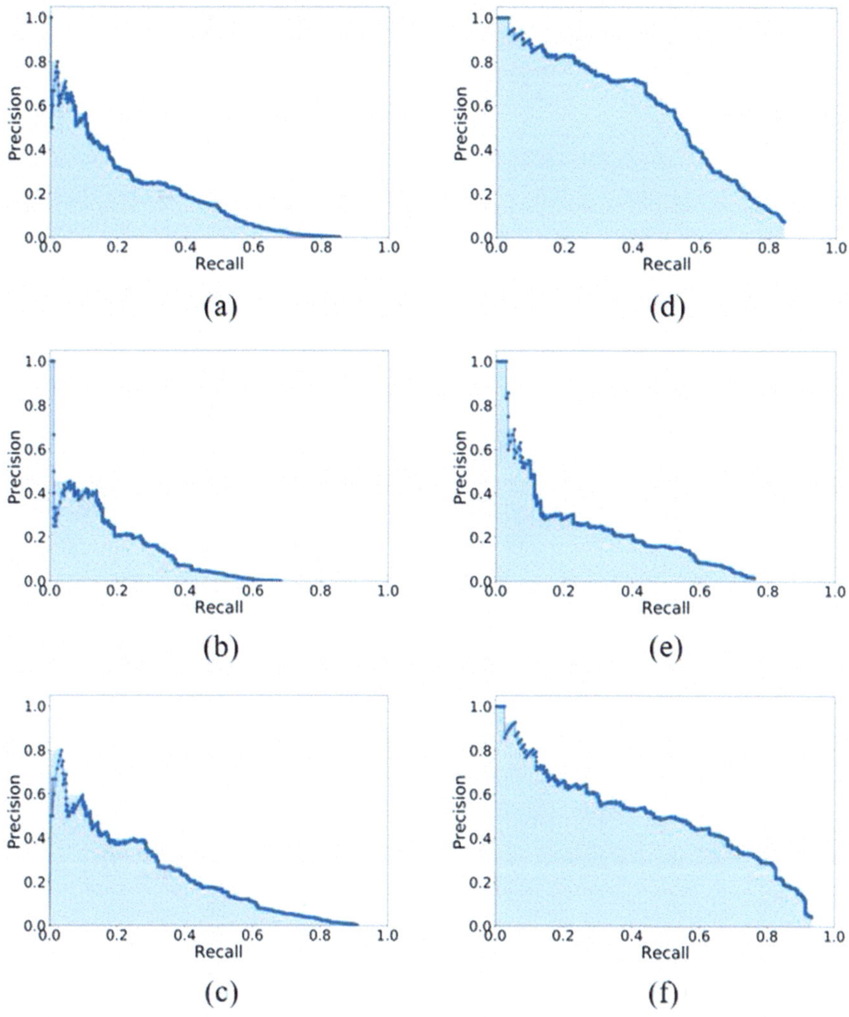

Fig. 21.10 Comparison of AP values for the two models

are overall higher than (a), (b), and (c), indicating that the YoloV3 model has better AP values.

The F1-Score threshold trend graph is shown in Fig. 21.11, which demonstrates the relationship between F1-Score and threshold. The (a), (b), and (c) represents the EfficientDetD0 model, and the (d), (e), and (f) represents the YoloV3 model. In the figure, the horizontal axis represents the threshold value and the vertical axis represents the F1-Score value. In the figure, we can see that as the threshold value increases, the F1-Score value shows an overall trend of first increasing and then decreasing. By comparing the two curves, we find that the F1-Score values of the

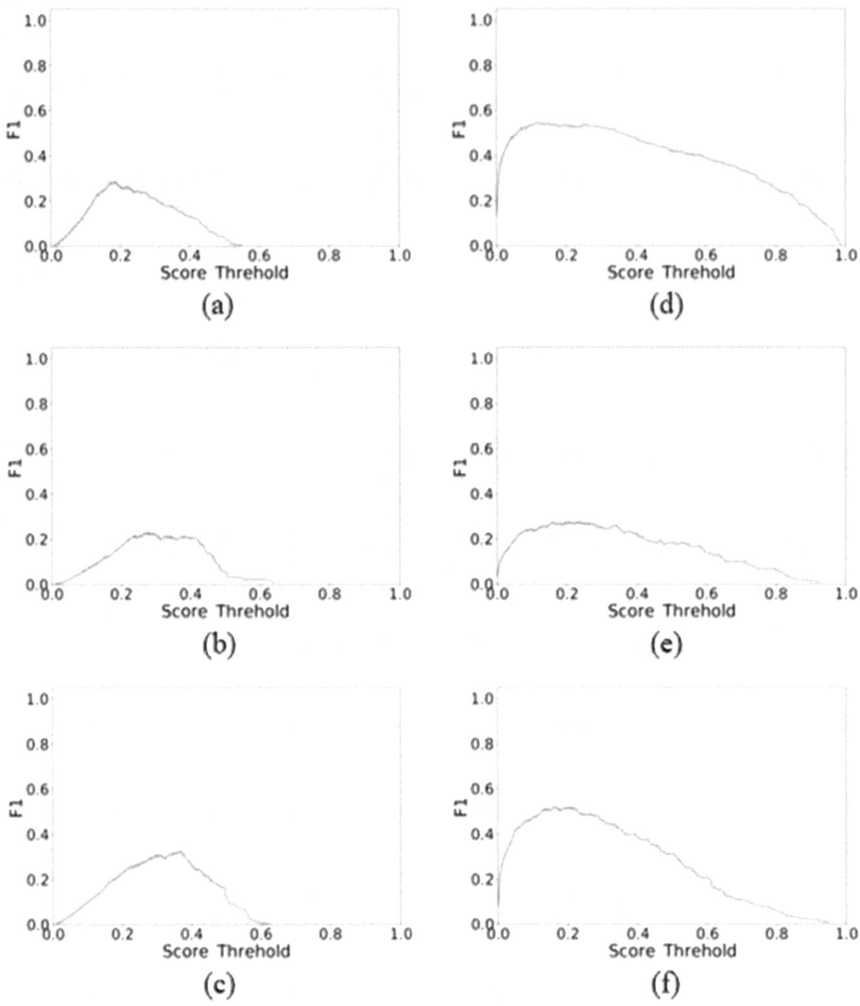

Fig. 21.11 Comparison of F1 values for the two models

YoloV3 model are higher than those of the EfficientDetD0 model for most of the threshold values, so we can conclude that the YoloV3 model has better performance in identifying the three categories of patterns compared to the EfficientDetD0 model.

Table 21.2 shows the data comparison of the two models for each index, the AP values of EfficientDetD0 for BZ, LX, and TW recognition are 19.07, 12.34, and 22.51% respectively, while the AP values of YoloV3 are 48.03, 17.91, and 47.79%, the three categories are higher than EfficientDetD0 respectively. The F1 values of EfficientDetD0 for BZ, LX, and TW are 0.04, 0.06, and 0.13, respectively, while those of YoloV3 are 0.38, 0.11, and 0.26, which were 0.34, 0.05, and 0.13 higher

Table 21.2 Experimental table comparing EfficientDetD0 with YoloV3

Model	Class	AP (%)	F1	Precision (%)	Recall (%)
EfficientDetD0	BZ	19.07	0.04	77.78	1.87
	LX	12.34	0.06	35.71	2.99
	TW	22.51	0.13	51.61	7.14
YoloV3	BZ	48.03	0.38	79.17	25.40
	LX	17.91	0.11	45.45	5.99
	TW	47.79	0.26	76.09	15.62

than those of EfficientDetD0, respectively. The Precision of EfficientDetD0 for BZ, LX, and TW is 77.78, 35.71, and 51.61%, respectively; the Precision of YoloV3 is 79.17, 45.45, and 76.09%, respectively, which was 1.39, 9.74, and 24.48% higher than EfficientDetD0. The Recall of EfficientDetD0 is 1.87, 2.99, and 7.14% for BZ, LX, and TW respectively, while the Recall of YoloV3 is 25.40, 5.99, and 15.62%, which is 23.53, 3.00, and 8.48% higher than EfficientDetD0 respectively. 8.48%. From the above data, it can be seen that the AP, F1, Precision, Recall of the YoloV3 network for the three kinds of recognition are significantly higher than that of the EfficientDetD0 algorithm, and there was a significant contrast in the precision rates, indicating that the YoloV3 algorithm used in this paper has a higher classification precision rate in the traditional costume patterns of the Atayal.

21.6 Conclusion

This paper focuses on the implementation of YoloV3 network based on the recognition of the traditional costume patterns of the Atayal, designed a comparison experiment of the network based on YoloV3 and EfficientDetD0, the experimental results show that YoloV3 for the recognition of the three kinds of The AP value is 28.96, 5.57, and 25.28% higher than EfficientDetD0; F1 value is 0.34, 0.05, and 0.13 higher than EfficientDetD0; Precision is 1.39, 9.74, and 24.48% higher than EfficientDetD0. Recall is 23.53, 3.00, and 8.48% higher than EfficientDetD0, respectively, and the overall data indicated that the YoloV3 network model selected based on this paper had better metrics.

Acknowledgments This study was funded by the Natural Science Foundation of Fujian Province (Project No.2020 J01286).

References

1. Guan, C., Qin, S., Long, Y.: Apparel-based deep learning system design for apparel style recommendation. Int. J. Cloth. Sci. Technol. **31**(3), 376–389 (2019)

2. Jo, J., Lee, S., Lee, C., et al.: Development of fashion product retrieval and recommendations model based on deep learning. Electronics. **9**(3), 1–12 (2020)
3. Liu, X., Sun, Y., Liu, Z., et al.: Learning diverse fashion collocation by neural graph filtering. IEEE Trans. Multimed. **23**, 1–8 (2020)
4. Quan, H., Li, S., Hu, J.: Product innovation design based on deep learning and Kansei engineering. Appl. Sci. **8**(12), 1–17 (2018)
5. Vijayaraj, A., Raj, V., Jebakumar, R., et al.: Deep learning image classification for fashion design. Wirel. Commun. Mob. Comput. **2022**, 7549397 (2022)
6. Zhang, C., Wu, S., Chen, J.: Identification of miao embroidery in Southeast Guizhou province of China based on convolution neural network. Autex Res. J. **21**(2), 198–206 (2021)
7. Chen, M., Wang, C., Qin, H.: Jointly learning shape descriptors and their correspondence via deep triplet CNNs. Comput. Aided Geom. Des. **62**, 192–205 (2018)
8. Lin, X., Rivenson, Y., Yardimci, N.T., et al.: All-optical machine learning using diffractive deep neural networks. Science. **361**(6406), 1–8 (2018)

Chapter 22
Design of Lifting Equipment of Wafer Unmanned Track Carrier

Yin-Hong Yao and Yi-Jui Chiu

Abstract The overhead hoist transport (OHT) in the automatic material handling system (AMHS) plays an important role in wafer fabrication. In this paper, a micro-hoist hoist driven by servo motor on OHT frame is designed. Firstly, the hoisting system layout is modeled, then the hoisting device and the clamping device are designed. Secondly, the modal analysis of the hoisting device is carried out by ANSYS simulation software. Thirdly, the hoisting device is accurately positioned and controlled by PCL. Finally, it is concluded that the strength and stiffness of the lifting equipment meet the functional requirements, and can achieve a stable lifting of 12.5 meters per second. At the same time, the lifting equipment and the OHT frame will not produce resonance, which meets the design purpose.

22.1 Introduction

Nowadays, the global semiconductor industry is increasingly competitive. China is in the middle of the semiconductor industry, which is the foundry of wafers. Optimizing OHT as a transport and loading tool is the way to improve plant efficiency. OHT pick-up methods and delivery routes should be continuously simplified, so the functions of OHT are increasingly centralized. Improving OHT body structure to achieve dynamic balance of vehicle flow is the development direction of OHT. The one-stop-arrival direct transport mode of OHT puts forward higher requirements for the lifting and clamping function of OHT. Wu et al. [1] established an integrated automatic material transportation system model and proposed a multi-objective scheduling method to improve wafer handling efficiency. Wu et al. [2] put forward the EMCM model and proved that the reference of this model can reduce the vehicle congestion rate. Wu et al. [3] proposed a dynamic scheduling method based on

Y.-H. Yao · Y.-J. Chiu (✉)
School of Mechanical and Automotive Engineering, Xiamen University of Technology, Xiamen, China
e-mail: chiuyijui@xmut.edu.cn

S.-P. Tseng et al. (eds.), *Advances in Intelligent Information Hiding and Multimedia Signal Processing, Volume 2*, Smart Innovation, Systems and Technologies 416, https://doi.org/10.1007/978-981-97-8760-9_22

fuzzy logic control, and proved through simulation that this method can improve the OHT handling efficiency in real time. In the study of composite dispatching rules (CDR), Qin et al. [4] proposed a CDR programming generator and proved that CDR can improve wafer throughput. In the problem of inconvenient manual material handling, Cai et al. [5] proposed a PLC-controlled handling equipment to improve production efficiency. In the study of PLC control application, Yun et al. [6] also designed a set of arm of a material handling manipulator. Yao [7] selected components according to the motion parameters of the manipulator to complete the control process of the manipulator. A new looper design idea was proposed by Jiang et al. [8] has the characteristics of more compact structure and lightweight in the study of winch design. Zhao [9] optimized the operation noise and drum string design to reduce noise and enhance equipment stability. Guo et al. [10] studied the resonance phenomenon in the work of the manipulator, used the finite element model to conduct modal analysis, obtained the natural frequency of the manipulator, solved the first six-order mode diagram, and analyzed the relationship between frequency and displacement. The above literature analyzed and designed the automatic material equipment, and designed and improved the lifting equipment and clamping device in OHT. The design of the lifting equipment is not only lightweight but also low noise and stable operation.

Based on the study of AMHS function and the design of automatic material handling equipment, this paper obtains the design idea of lifting equipment on OHT, and designs a set of lifting equipment composed of micro-hoist and clamping mechanism. Compared with the traditional hoist, less structural parts such as the reducer are removed, and the lifting equipment is driven by servo motor to achieve accurate positioning control.

22.2 Structural Design of Lifting Equipment

According to the structure layout of the lifting equipment in the AMHS system as shown in Fig. 22.1. On the basis that the internal structure of OHT is compact and the FOUP lifting accuracy is guaranteed, the OHT structure is modeled, as shown in Fig. 22.2. Because the servo motor positioning accuracy can reach the um level, this can meet the accuracy of FOUP clamping during the lifting process, and the servo motor equipped with the servo drive PLC to help achieve a number of different movement requirements. Hoist in all kinds of lifting equipment has their figure, because their structure is simple and reliable. So I designed a lifting device based on the prototype of the hoist. The design structure of lifting equipment is shown in Fig. 22.3.

Expression of power required in full load working condition:

$$P = \frac{F \times V}{1000} \tag{22.1}$$

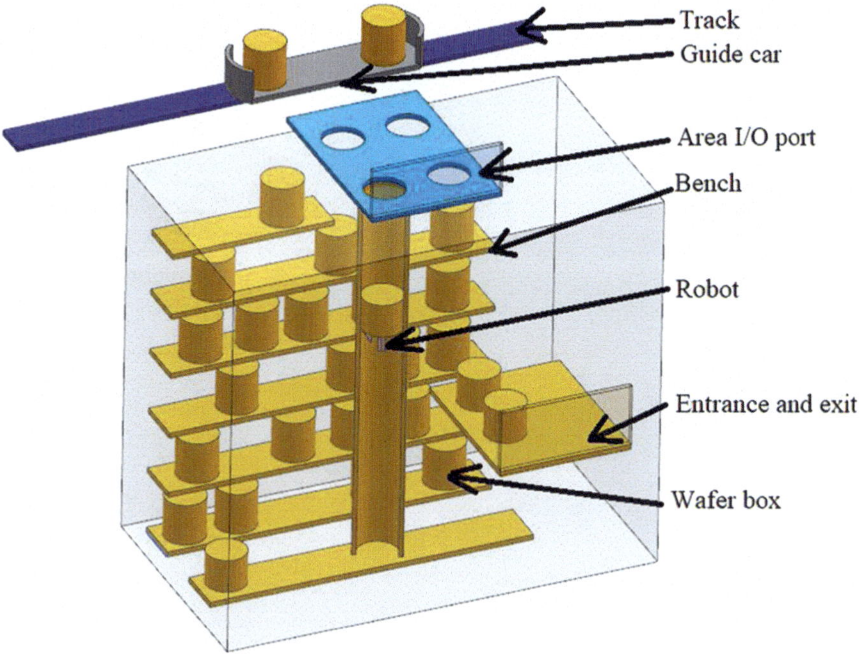

Fig. 22.1 Structure layout of lifting system in AMHS system

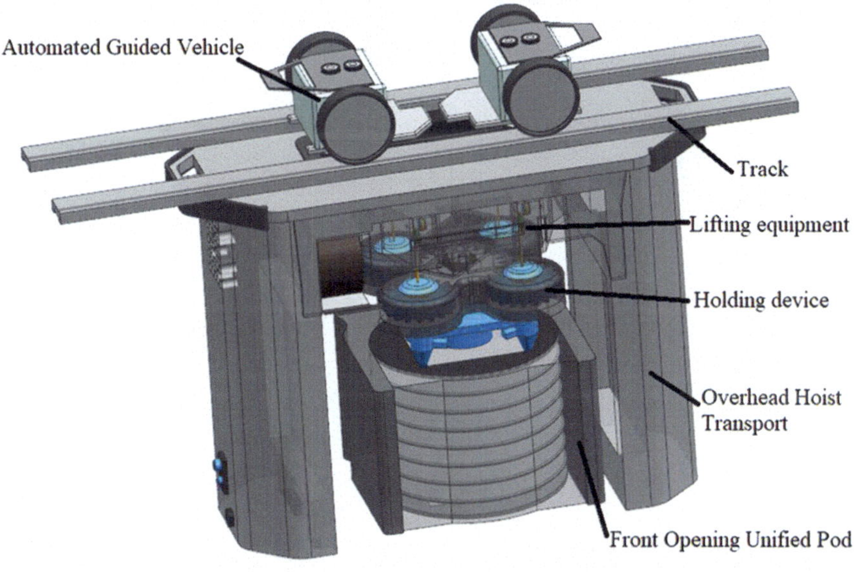

Fig. 22.2 OHT structure

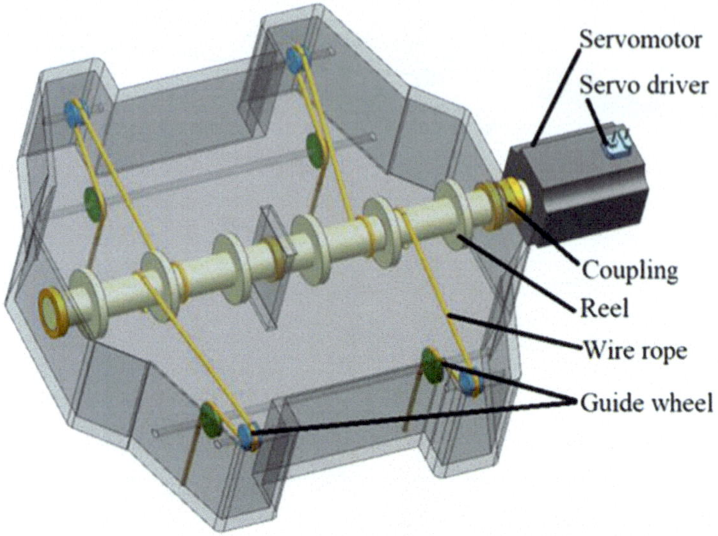

Fig. 22.3 Lifting equipment structure

where P is the required power (kW), F is full load tension (N), and V is full load lifting speed (m/s).

Wire rope minimum diameter calculation expression:

$$d_{min} = c\sqrt{F} \tag{22.2}$$

where d_{min} is the minimum diameter of the wire rope, c is the selection factor$\left(mm/\sqrt{N}\right)$.

Drum pitch diameter calculation expression:

$$D \geq K_r d \tag{22.3}$$

where K_r is the ratio of the drum to the wire rope, d is the diameter of the wire rope. Wire rope diameter calculation expression:

$$D_0 \geq K_r d \tag{22.4}$$

$$D_0 = K_r d - d \tag{22.5}$$

where $D0$ is the diameter of the wire rope.

$$B_t < 3D_0 \tag{22.6}$$

$$D_0 \geq K_r d \tag{22.7}$$

Calculation expression of side plate diameter:

$$Dk \geq Ds + 4d = D0 + (2S - 1)\,d + 4d \tag{22.8}$$

where Ds is the core diameter of the outermost wire rope, S is the number of winding layers of the wire rope.

Drum wall thickness calculation expression:

$$\delta = 0.02D + (6 \sim 10) \tag{22.9}$$

The torque calculation expression of the coupling:

$$T_c = K \times \frac{9550P}{n} \tag{22.10}$$

where T is torque, P is motor power, and n is motor speed.

According to GB/T 5014–2003, type LXZ2 elastic pin coupling with brake wheel and type-specific torque can be found.

22.3 Design of Clamping Equipment

The clamping device designed is a flange four-notch holding head with a stepping motor as a power source. The structure of the clamping equipment is simple as shown in Fig. 22.4. When clamping, the motor drives the driving gear to rotate, and the driving gear drives the driven gear to rotate 180°. The notch below the driven gear matches the groove on the FOUP to realize the clamping. When released, the motor again rotates the driven gear 180° to release the FOUP. The device almost does not need to do work, so a low-power motor is used. The power is set to 0.1 kW. Transmission ratio i is 1.353, one-way operation, stable load. 45 steel is used for small wheel, tempered with hardness of 229–286 HBS. The 45 steel is used for large wheel, normalizing with hardness of 169–217 HBS. Because the hardness is less than 350 HBS, it is a soft tooth surface gear. It is designed according to the contact fatigue strength, and then the bending fatigue strength can be checked to complete the design requirements.

$$T_1 = \frac{9500 \times 10^6 \times P}{n_1} \tag{22.11}$$

The load coefficient K is obtained by referring to the table and K is 1.1. From the number of gear calculation expression:

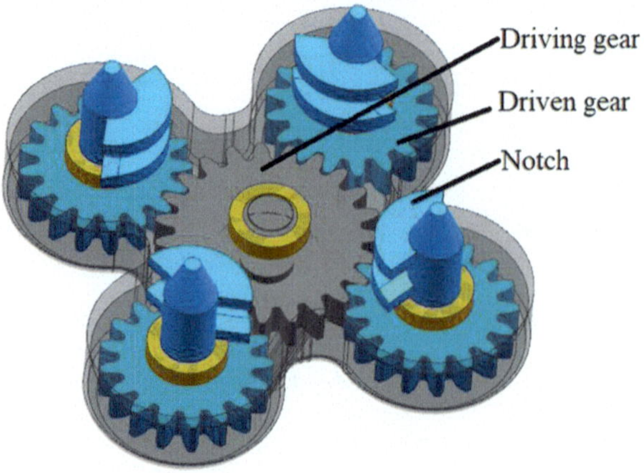

Fig. 22.4 Design of clamping equipment Expression of torque transmitted from the driving wheel

$$z_2 = i \cdot z_1 \tag{22.12}$$

The gear transmission is arranged symmetrically, and it is a soft-tooth surface gear.

$$\psi_d = -1$$

Look up the table to get the values of $^{\sigma}Hlim1$ and $^{\sigma}Hlim2$.

$$N_1 = 60N \cdot J \cdot L_h \tag{22.13}$$

$$N_2 = \frac{N_1}{i} \tag{22.14}$$

$$[\sigma_H]_1 = \frac{Z_{N1} \cdot \sigma_{Hlim1}}{S_h} \tag{22.15}$$

$$[\sigma_H]_2 = \frac{Z_{N1} \cdot \sigma_{Hlim1}}{S_h} \tag{22.16}$$

$$d_1 \geq \sqrt{\frac{2K \times T_1 \times (u+1)}{\psi_d \cdot u} \left(\frac{Z_E Z_H Z_\varepsilon Z_\beta}{\sigma_H} \right)^2} \tag{22.17}$$

$$m = \frac{d_1}{z_1} \tag{22.18}$$

Driving wheel diameter expression:

$$d_1 = mz_1 \tag{22.19}$$

Driven wheel diameter expression:

$$d_2 = mz_2 \tag{22.20}$$

Driven gear tooth width expression:

$$b = \psi_d \times d_1 \tag{22.21}$$

Drive gear tooth width expression:

$$b_1 = b_2 + 3 \tag{22.21}$$

Center distance:

$$a = m\,(z_1 + z_2) \tag{22.22}$$

The tooth profile coefficient *YN1*, *YN2*, stress correction coefficient *YS1*, *YS2*, and allowable bending stress *Flim*1 and *Flim*2 were obtained from the material.

The bending stress limit expression of gear is as follows:

$$[\sigma_F] = \frac{Y_{N1} \cdot \sigma_{Flim}}{S_F} \tag{22.24}$$

Bending stress expression of driving gear:

$$\sigma_{F1} = \frac{2K \times T_1}{bmz} Y_{F1} Y_{F2} \tag{22.25}$$

Bending stress expression of driven gear:

$$\sigma_{F2} = \sigma_{F1} \frac{Y_{F2} Y_{s2}}{Y_{F1} Y_{S2}} \tag{22.26}$$

The results show that the root bending fatigue is qualified and meets the design requirements.

22.4 Modal Analysis

OHT modeling can be obtained from the structural design part of OHT. In lifting equipment, it can be seen that the maximum load of fully loaded wafers is 250 N,

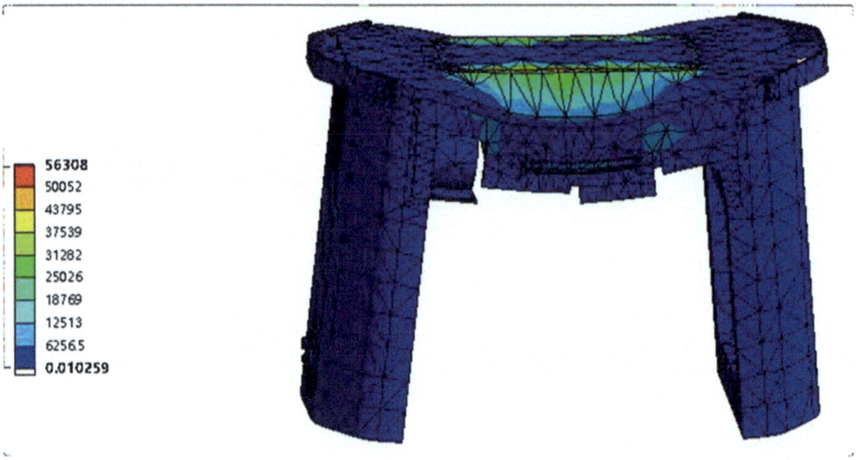

Fig. 22.5 Overall deformation diagram of OHT

and the self-weight of OHT is about 750 N. Therefore, the chassis load of the guide car was calibrated to 1000 N for analysis.

ANSYS workbench is used in this analysis. It is the only co-simulation platform in the world, which can perfectly combine simulation technology and simulation data. ANSYS workbench can carry out statical analysis on the model of mechanical mechanism and suggest areas that can be optimized. At the same time, the advantages and disadvantages of different structures can be compared [11, 12]. The simulation of multiple physical fields also meets the goal pursued by modern CAE software. In workbench analysis, the model structure in UG is first derived into X_T file type. Then the geometric structure is inserted into the structure modeling in the component system of workbench 2022 R1 of ANSYS neutron program, and then the static structure of the analysis system is analyzed statically. The equal scale used in the previous structural modeling is to ensure the return of the original scale and the connection of the model parts, so as to avoid the failure to release the stress analysis, deformation and other force cloud map due to the error of structural strength.

After ANSYS analysis, the stress nephogram of OHT is shown in Fig. 22.5. The deformation diagram of OHT is shown in Fig. 22.6. And the grid diagram of OHT is shown in Fig. 22.7.

It can be seen from Figs. 22.5, 22.6 and 22.7 that in the case of structural steel, the maximum total deformation is 3.570×10^{-9} mm, and the maximum stress is less than 5.005×10^5 Pa < [σ] =120 MPa, so it meets the design of structural strength.

It can be seen from Fig. 22.8, 22.9 and 22.10 that in the case of the 45 steel, the maximum total deformation is 3.105×10^{-7}mm in Fig. 22.9. And the maximum stress shown in Fig. 22.10 is less than the allowable stress

The prestressed modal analysis of the clamping device was carried out to obtain the modal diagram of order one to six as shown in Fig. 22.11.

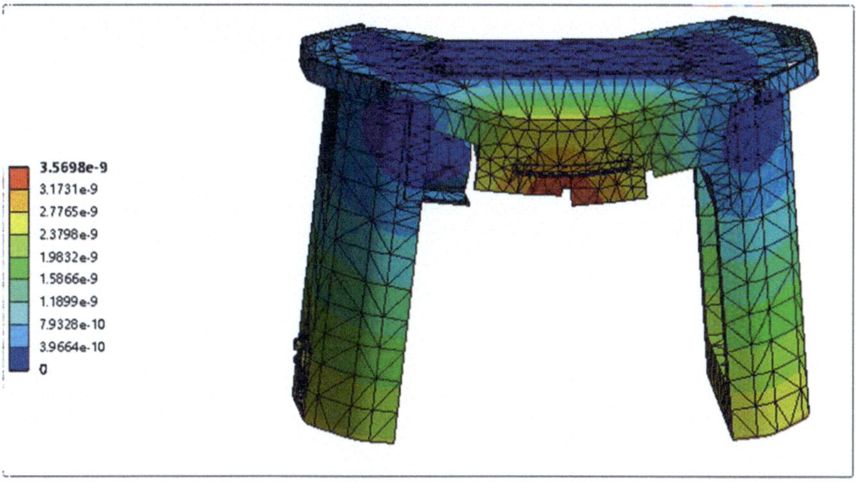

Fig. 22.6 Equal-effect diagram of OHT

Fig. 22.7 Grid diagram of OHT

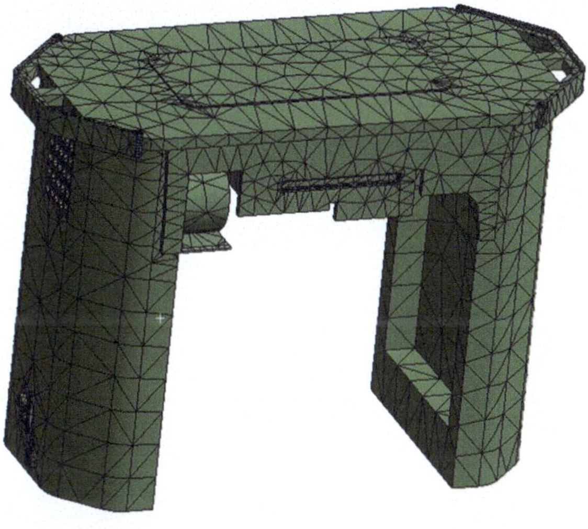

The exciting vibration of the clamping device in the working process may come from the working vibration of the motor, the working vibration of the hoist, and the vibration of the wire rope running. As shown in Table 22.1, the OHT first-order natural frequency of the lifting equipment in workbench modal analysis is 1089.0.5 Hz. When the wafer unmanned rail carrier is lifting smoothly, the excitation frequency of the lifting equipment excitation source can be seen in the range of 3003.9 to 3487.7 Hz as shown in Table 22.2. The modes of 1 to 5 of the clamping device are staggered with the first natural frequency of OHT, which can avoid the occurrence of resonance. In actual production, it is also necessary to

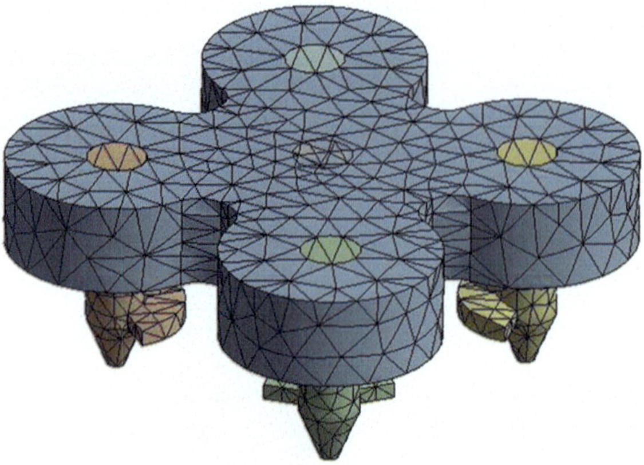

Fig. 22.8 Grid diagram of clamping device

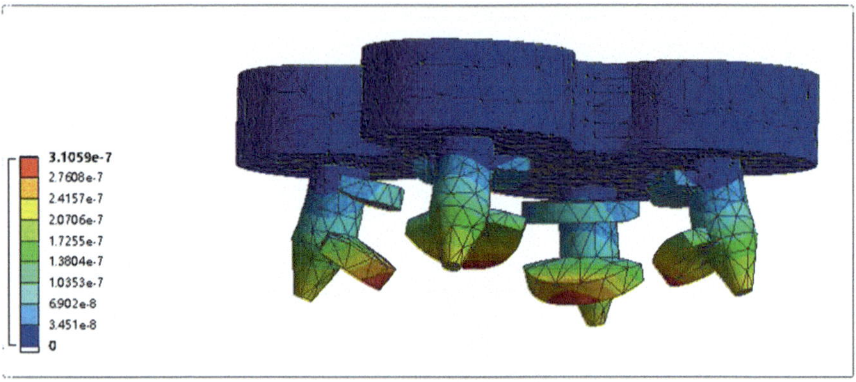

Fig. 22.9 Overall deformation diagram

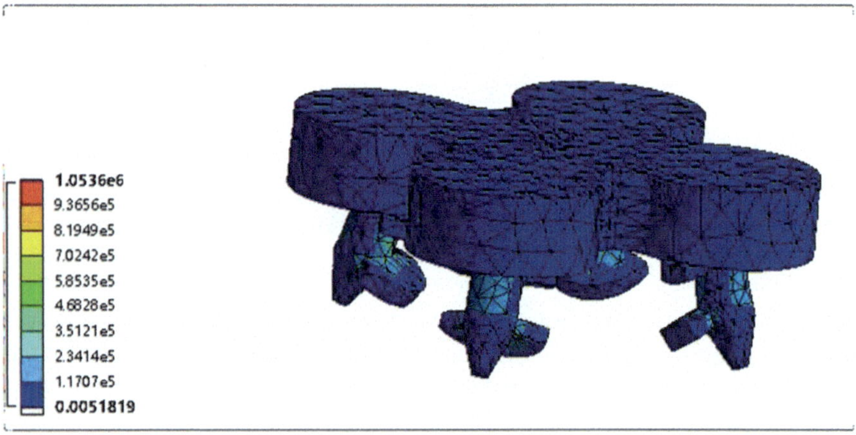

Fig. 22.10 Equal-effect diagram

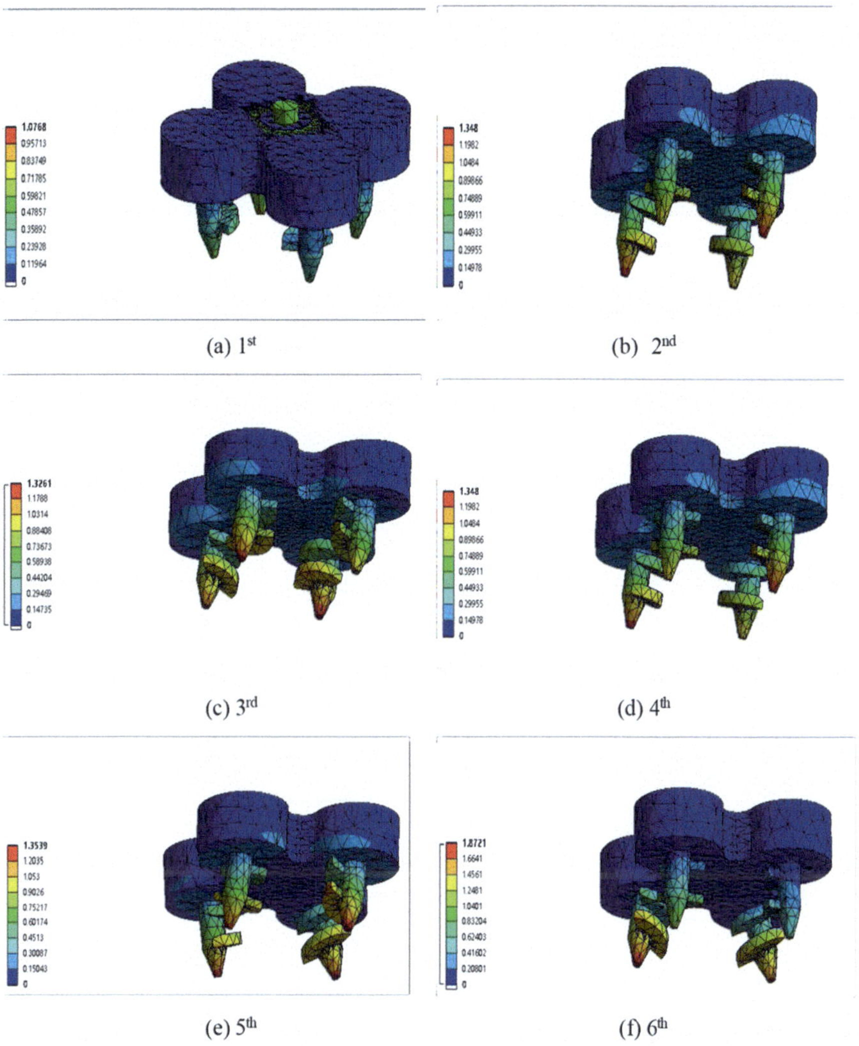

Fig. 22.11 Modal diagram of the first to sixth order of the frame. (**a**) 1st. (**b**) 2nd. (**c**) 3rd. (**d**) 4th. (**e**) 5th. (**f**) 6th

consider the wafer box used for wafer transportation, the vibration during contact, and so on, which needs to be analyzed on a case-by-case basis.

Table 22.1 Natural frequency of order 6 of OHT

Mode	Frequency (Hz)
1	1089.5
2	1407.3
3	1863.0
4	2028.1
5	2171.3
6	3533.4

Table 22.2 The frequencies and mode of the vehicle body

Mode	Frequency (Hz)	Maximum deformation (mm)	Mode characteristics
1	3003.9	0.957	It vibrates up and down around the center
2	3035.7	1.048	Side-to-side vibration
3	3037.5	1.179	Fore and aft vibration
4	3128.4	0.780	Center up and down vibration
5	3222.7	1.203	All sides vibrate to the center
6	3487.7	1.664	The center vibrates in all directions

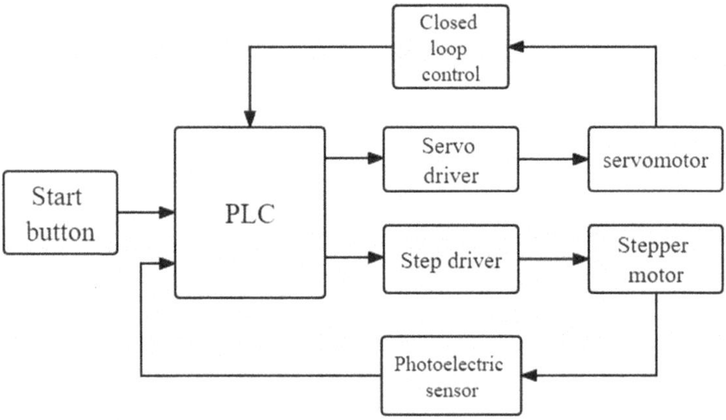

Fig. 22.12 Control diagram

22.5 Drive Control and Conclusion

OHT's lifting equipment consists of a 4-wire rope lifting manipulator and a motor. The servo motor is controlled by PLC. The control idea is shown in Fig. 22.12. Because PLC has few functions and environmental limitations, the DVP40ES200 of the Delta small controller DVP-ES2 series controls the servo motor forward rotation to achieve the lifting function. The clamping device is used to achieve a lifting function. The PLC ladder diagram program is shown in Fig. 22.13.

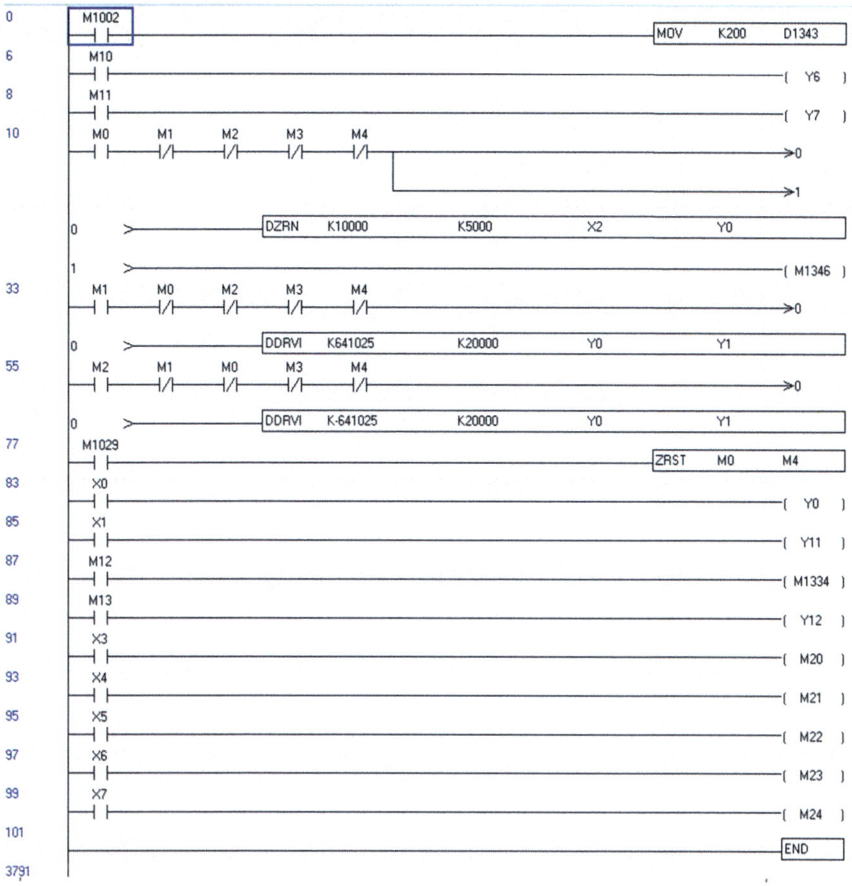

Fig. 22.13 Ladder diagram program

The research purpose of this topic is to design and study the lifting equipment of wafer unmanned rail carrier under full load condition, and the results are as follows:

1. The stress analysis results in Ansys workbench show that when the original model is bent at full load, the maximum stress in the frame of OHT loading lifting equipment is 5.525Mpa, which does not reach the yield strength of the frame material. The deformation analysis results show that the maximum deformation displacement value is 0.990 mm, which is small and can be ignored. When the original model is running at full load, the stress analysis results of the clamping device show that the maximum stress is 3.512Mpa, which obviously cannot reach the yield strength of the material. The deformation analysis results show that the maximum deformation displacement is 3.105×10^{-7} mm, which is a small value and can be ignored. According to the analysis results, the sleeve lifting

equipment is optimized. Re-analysis of the working condition shows that the maximum stress is 6.892Mpa, which is still not enough to reach the yield limit of the frame material and meet the requirements. The maximum deformation displacement is 0.147 mm and 0.734 mm respectively, and the deformation is small.

2. The simulation results show that the natural frequency range of the OHT frame is 1089.5 Hz, while the vibration generated by the excitation of the lifting equipment during the driving process is 3003.9–3487.7 Hz. Under normal working conditions, the lifting equipment is unlikely to damage the body structure due to resonance with the OHT during lifting.

Acknowledgments These entire study works are sustained by Fujian Industrial Technology Development and Application Project NO.2021I0024.

References

1. Wu, L.H., Song, H.J., Zhang, Z.W., et al.: AMHS multi-objective scheduling method for wafer manufacturing for rush orders. Mod. Mach. Tool Auto. Manuf. Techniq. **1**, 155–159 (2020)
2. Wu, L.H., Zhang, Z.W., Zhang, J., et al.: A performance model of automated material handling systems with double closed-loops and shortcuts in 300 mm semiconductor wafer fabrication systems. J. Manuf. Syst. **58**, 316–334 (2021)
3. Wu, L.H., Song, H.J., Zhang, J.X.: Integrated AMHS dynamic scheduling method for wafer fabrication based on fuzzy logic. Mod. Mach. Tool Auto. Manuf. Techniq. **05**, 152–155 (2019)
4. Qin, W., Zhang, J., Sun, Y.B.: Multiple-objective scheduling for interbay AMHS by using genetic-programming-based composite dispatching rules generator. Comput. Ind. **64**, 694–707 (2013)
5. Cai, H.M., Hao, T.H., Liu, M.Z., et al.: Design and research of handling equipment based on PLC control. Autom. Technol. Appl. **37**(12), 32–34 (2018)
6. Yun, L.H., Lu, Y.: Design of mechanical arm for loading and unloading materials based on PLC. China Metalform. Equip. Manuf. Technol. **57**(6), 52–54 (2022)
7. Yao, F.-J.: Design and PLC control of manipulator. Mech. Manag. Dev. **239**(3), 80–82 (2023)
8. Jiang, R.C., Liu, W.X.: Design of looper hoist. First Heavy Technol. **168**(06), 5–9 (2015)
9. Zhao, D.X.: Optimization design of wire rope winch. China Equip. Eng. **468**(06), 137–138 (2021)
10. Guo, B.T., Dai, G.H.: Structural dynamics analysis of gantry manipulator based on Ansys workbench. Shanxi coking. Coal Sci. Technol. **46**(11), 21–48 (2022)
11. Fan, K.: Application of finite element analysis in mechanical design and manufacture. Eng. Test. **61**(2), 86–89 (2021)
12. He, Y.M., Hao, A.B.: Calculation of gear meshing stiffness based on ANSYS workbench and dynamic simulation. J. Shenyang Univ. Technol. **42**(2), 191–196 (2020)

Chapter 23
Improving the Teaching Quality of International Students via Intercultural Teachers in Higher Education of Information Technology

Xiaohong Quan, Shih-Pang Tseng, and Wenqiang Guan

Abstract The significance of teaching for international students in higher education institutions of information technology is analyzed, and the problems and challenges faced by higher education institutions of information technology are listed. The advantages of intercultural teachers in cross-cultural teaching are expounded, and the feasible measures and evaluation system are put forward to promote the intercultural teachers to participate in the teaching of international students. It is suggested to strengthen the leading role of intercultural teachers in cross-cultural teaching, and to promote the improvement of the teaching quality of international students in higher education institutions of information technology.

23.1 Introduction

The development of information technology (IT) efficaciously gave an impetus to the globalization. In the same way, the globalization accelerated the development of IT [6]. Therefore, the globalization of IT education has become an important and urgent high education issue. Relative to the general universities, the higher education institutions of information technology focus on the IT and related domains. The higher education institutions of IT are more sensitive about the globalization issue rather than the general universities [1]. Furthermore, the international students and the intercultural teachers are necessary to construct a globalized high education environment. In this paper, we studied the influence of intercultural teachers on the international students in the higher education institutions of information technology from various viewpoints.

X. Quan (✉) · S.-P. Tseng · W. Guan
School of Software and Big Data, Changzhou College of Information Technology, Changzhou, China
e-mail: quanxiaohong@czcit.edu.cn; tsengshihpang@czcit.edu.cn; guanwenqiang@czcit.edu.cn

S.-P. Tseng et al. (eds.), *Advances in Intelligent Information Hiding and Multimedia Signal Processing, Volume 2*, Smart Innovation, Systems and Technologies 416, https://doi.org/10.1007/978-981-97-8760-9_23

23.1.1 Promote International Friendly Relations and Cultural Exchanges

International students' study experience not only enables them to gain professional knowledge, but also enables them to build deep friendships with domestic students and faculty. Such international friendly relations and cultural exchanges help to enhance understanding, respect, and friendship among countries, break down cultural barriers, and promote cross-cultural exchanges and cooperation [3].

23.1.2 Strengthen the Training of Information Technology Talents

The higher education institutions of information technology usually focus on practical skills in information technology, which satisfied the needs of globalization. By providing international students with professional education in information technology, the higher education institutions of information technology can cultivate more talents with practical skills and application abilities, which will help meet the development needs of the global countries.

23.1.3 Promote Economic Cooperation and Development

The globalization aims to promote cross-border cooperation and economic development. The international students not only contribute to the development of their own countries, but also serve as a link to promote cooperation and exchanges among countries. Their learning experience and skills in information technology can provide strong support for their country's industrial and economic development.

23.2 Challenges in Teaching International Students in Higher Education Institutions of Information Technology

23.2.1 Language Barrier

International students are from different countries and regions, where their first language may not be the medium of instruction in the institutions, leading to difficulties in classroom understanding and expression. If the institutions cannot provide effective language support and training, international students may have language communication barriers [2], affecting their learning effect and participation.

23.2.2 Faculty

The teaching level and cross-cultural teaching ability of teachers are very important to the teaching of international students. An international teaching team with rich cross-cultural teaching experience and excellent professional knowledge and skills is the guarantee of the teaching quality of international students and can make international students obtain satisfactory learning experience. In addition, an excellent international teacher team should not only teach professional knowledge, but also be good at establishing a good teacher-student relationship with international students, so that they can gain a sense of belonging and security in the school.

23.2.3 Curriculum Design and Teaching Methods

International students come from different cultural backgrounds and education systems, and their academic needs may differ from those of local students. Therefore, curriculum design needs to be flexible to ensure that material content and course structure can match the background and academic goals of international students. International students have different language ability, learning style, and ability level. Teaching methods thereby should adapt to these differences and adopt diversified teaching strategies to effectively motivate the enthusiasm of international students and achieve satisfied learning results.

23.2.4 Teaching Resources

Different countries and regions have different levels of teaching resources and equipment. Therefore, international students need to adapt to the new teaching environment. Whether higher education institutions of information technology can provide rich learning resources is very important to the learning experience and academic development of international students. Teaching strategies should consider how to maximize the use of existing teaching resources to achieve better teaching results.

23.2.5 Cultural Adaptation and Support

International students face certain challenges when adapting to the new learning and living environment. Whether the higher education institution can provide adequate cultural adaptation and psychological support for international students, including guiding international students to adapt to the new environment, providing psychological counseling, and helping them solve life problems. Whether international students have enough opportunities to communicate and integrate with local students and communities is an important factor that affects the quality of teaching for international students.

23.3 Advantages of Intercultural Teachers in Teaching of International Students

23.3.1 Advantages of Language and Cross-cultural Communication

Based on overseas study experience and multi-cultural experience, intercultural teachers are not only fluent in foreign language communication, but also have a deep understanding of the differences and commonalities between different cultures, which enables them to provide targeted guidance to international students, help them overcome language barriers, and improve their listening, speaking, reading, and writing skills. At the same time, intercultural teachers can better understand the cultural background and way of thinking of international students, as well as avoid possible misunderstandings and conflicts in teaching. They are able to create a positive cross-cultural learning atmosphere in the classroom and promote the integration of international students with the local culture [4].

23.3.2 Advantages of International Teaching Resources

By making use of the international background, intercultural teachers can actively introduce international teaching resources and integrate them into the teaching design of international students. Through the introduction of curriculum, exchange programs, international teaching materials, journals, etc., international students can access the frontier academic knowledge and development trends in the world and expand the depth and breadth of students' knowledge. In addition, intercultural teachers are familiar with modern technological means, and can also develop their own teaching resources, provide online learning platforms, virtual laboratories, etc., to further enrich the learning experience of international students.

23.3.3 Advantages of Diversified Teaching Methods

Intercultural teachers who have studied abroad are familiar with educational concepts and methods in different countries and culture. They can flexibly use advanced

Table 23.1 Diversified teaching methods and their applicable scenarios

Number	Teaching method	Applicable scenario
1	Lecture method	The imparting of theoretical knowledge
2	Multimedia teaching method	The teaching of abstract concepts or complex processes, such as mathematical formulas or practical cases
3	Problem-inquiry method	Deepening understanding of a particular issue through in-depth research by students, as well as conducting inquiry-based learning activities such as scientific experiments
4	Training-practice method	Development of practical skills and application of theoretical knowledge to enhance problem-solving abilities and gain practical experience
5	Demonstration teaching method	Development of practical skills for students in simulated or actual work environments to enhance their ability to perform tasks independently
6	Simulation teaching method	Simulated or actual work scenarios for students to gain knowledge and skills through practical application. This approach allows students to apply their theoretical knowledge in a controlled environment
7	Project-based teaching method	The acquisition of knowledge and skills through active participation in completing a real-world project that requires practical application of theoretical knowledge

teaching strategies and means to integrate innovative and diverse teaching methods into curriculum teaching. Table 23.1 shows diversified teaching methods and their applicable scenarios. With the help of diversified teaching methods, intercultural teachers can fully stimulate students' learning interest and provide personalized learning guidance to meet the needs of different students.

23.3.4 Advantages of International Cooperation and Career Development

Intercultural teachers who have studied abroad usually have established cooperative relations with foreign universities and institutions, which is conducive to conduct cross-border cooperation projects, leading students to actively participate in international academic seminars, competitions, and projects, providing students with a learning experience that is in line with international standards, cultivating students' thinking mode of transnational cooperation and improving students' international communication skills and academic communication ability. Intercultural teachers who have accumulated some work experience and career resources abroad and have a better understanding of the international job market can provide more career guidance and advice for international students, as well as prepare them for their future career development in an international environment.

23.4 Measures to Promote Participation of Intercultural Teachers in Teaching of International Students

The higher education institutions of information technology should take various measures to promote the participation of intercultural teachers to improve their level of internationalization [5]. Here are some possible measures:

23.4.1 Establish an International Teaching Team

As an important resource of international education, the intercultural teachers should be included in the professional teaching team. Higher education institutions of information technology should attract more intercultural teachers with teaching experience and professional skills to participate in the education of international students through recruitment and training, integrate them with domestic teachers to form cross-field, cross-cultural, and international teams, and promote the complementary and cooperation among team members so as to improve the internationalization level in higher education institutions of information technology.

23.4.2 Develop International Curriculum

Intercultural teachers who have studied abroad play an important role in the development of international education plans. They can integrate their own experience of studying and aboard working into the teaching design, comprehensively consider the language level, curriculum needs, and career development goals of international students, and formulate international teaching plans that are close to the international trend and needs of international students.

23.4.3 Establish a Tutor System for International Students

A tutor system for international students should be established with intercultural teachers serving as tutors. The international background of intercultural teachers enables them to better understand the needs of international students. They provide guidance and support to international students through personalized counseling, including academic planning, career development, cross-cultural communication, and research trends, to help international students integrate into the new environment more smoothly and expand their international horizons to achieve academic and career goals.

23.4.4 Carry Out International Exchange Activities

International exchanges and cooperation between universities, research institutions, and enterprises can be strengthened with the help of intercultural teachers through i.e., international seminars, academic lectures, cultural festivals, international volunteer projects, scientific research projects, etc. Through these online or offline activity platforms, more international learning and practice opportunities will be provided for international students and intercultural teachers to enrich their academic and career development experiences.

23.4.5 Provide Training and Support

Providing professional training, including language training, education laws and regulations training, etc., can help intercultural teachers adapt to the domestic educational environment. Higher education institutions can also provide more resource support to help intercultural teachers better prepare for teaching. It is a good practice to encourage intercultural teachers to engage in international education research, and support them to publish papers in international journals. It is also important to create an international student support system to provide better academic and life support to international students, including acculturation, academic counseling, and mental health support.

23.4.6 Establish Continuous Improvement and Incentive Mechanism

Through regular teaching evaluation and teaching feedback, international students and intercultural teachers can participate in the continuous improvement of the teaching quality of international students. Higher education institutions should establish an incentive mechanism to encourage intercultural teachers to achieve excellent results in the teaching of international students, including teaching results, scientific research results, student satisfaction, and other aspects.

23.5 Evaluation of Teaching Ability of Intercultural teachers

Higher education institutions of information technology should actively support and encourage intercultural teachers to participate in the education of international students and establish an effective evaluation system to ensure the quality of the teaching. Figure 23.1 shows the teaching quality evaluation system for intercultural

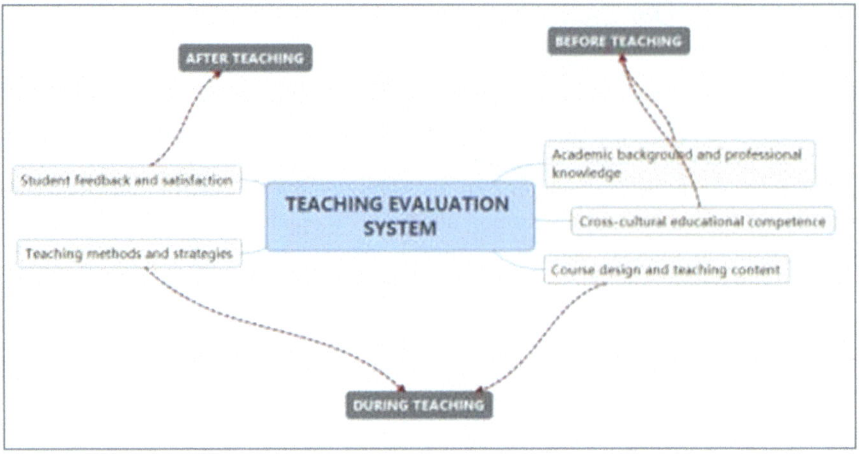

Fig. 23.1 Teaching evaluation system for intercultural teachers

teachers. Evaluation of teaching ability of intercultural teachers is not only beneficial to the international students and intercultural teachers, but also helps the higher education institutions of information technology to better fulfill their international education mission.

23.5.1 Assessment of Academic Background and Professional Knowledge

Academic background and professional knowledge assessment includes assessing the academic background, research experience and breadth and depth of expertise of intercultural teachers. The academic background of intercultural teachers should be consistent with the course they are teaching to ensure that they are able to provide high-quality international teaching. It is important to ensure that intercultural teachers have sufficient professional knowledge to teach the relevant curriculum.

23.5.2 Assessment of Cross-cultural Educational Competence

Intercultural teachers should have sufficient intercultural teaching experience to help international students adapt to cross-cultural learning and communication. This requires intercultural teachers to choose appropriate teaching methods and means according to the cultural background and language level of international students, as well as be able to carry out teaching effectively in different cultural atmospheres. Intercultural teachers should help international students to establish an open, tolerant, and respectful attitude towards other cultures, as well as properly solve difficulties encountered due to cultural differences.

23.5.3 Assessment of Course Design and Teaching Content

This part covers the intercultural teachers' curriculum design and teaching content selection. Intercultural teachers should consider the overall design of the curriculum, including curriculum objectives, curriculum outline, process evaluation, and textbook selection. Curriculum design should be consistent with international standards and trends to meet the academic needs of international students and have practical application value.

23.5.4 Assessment of Teaching Methods and Strategies

This part covers intercultural teachers' teaching methods, teaching strategies, and the utilization of teaching resources. Assessments should focus on whether intercultural teachers are able to use a variety of teaching methods, including interactive classes, case studies, and group discussions, to meet the learning styles and needs from different students. It is important to evaluate whether intercultural teachers are using updated teaching materials and resources, as well as whether they are adopting innovative teaching tools and technologies.

23.5.5 Assessment of Student Feedback

Student satisfaction and feedback can provide important information about teaching quality and help improve teaching. Higher education institutions should integrate the teaching quality assessment of international students into the overall teaching quality evaluation system of the institution, so that the whole teaching process of international students can be recorded, monitored, evaluated, and traceable. A multi-dimensional evaluation system can help intercultural teachers to find their shortcomings timely and improve their teaching quality.

23.6 Conclusion

In the context of globalization, high-quality teaching for international students is crucial for higher education institutions to improve their education level and enhance their international influence and competitiveness. Due to their unique cross-cultural advantages, intercultural teachers have great value in teaching for international students. Higher education institutions of information technology should take a series of measures to strengthen the participation and leading role of intercultural teachers in the teaching of international students. It is also worth to note that, to realize their maximum value in the teaching of international students, intercultural teachers still need continuous training and resource support that depends on the joint efforts of the government, higher education institutions, and all sectors of society to build a more open and diversified platform where intercultural teachers can play a greater role in the teaching of international students.

Acknowledgments This work is partially supported by Education and Teaching Reform Project of Changzhou College of Information Technology in 2023: Study on Strategies for Improving Teaching Quality for International Students (2023CXJG20), United Front Research Project of Changzhou College of Information Technology in 2023: Study on the Political Concept and Political Participation of Returnees (2023CXTZ03), The Teaching Innovation Team Project of Changzhou College of Information Technology in Software Technology (SG05020301050701), and New Form Integrated Textbook Construction Project of Changzhou College of Information Technology (SG04010303005).

References

1. Hao, P.: Research on the education and teaching of Higher Vocational College Students under the background of One Belt and One Road: a case study of Changzhou Textile institute. J. S. Edu. **3**, 122–134 (2019)
2. Zhang, R., et al.: Discussion on the teaching situation and coping strategies of international students in higher vocational colleges – a case study of Nantong Vocational University. J. Nantong Vocational University. **21**(03), 42–45 (2017)
3. Wang, Q.: A Preliminary study on the recruitment of international students in higher vocational colleges from the perspective of the belt and road initiative. Overseas English. **2**, 98–99 (2019)
4. Xu, L.: Career value return of returning college teachers in reverse culture shock. New West. **17**, 122 (2016)
5. Xu, Y.: Characteristics and management countermeasures of the new generation of young teachers returning from abroad in Xuyi University. Sci. Adv. **13**, 56–57 (2011)
6. Qadri, B., Bhat, M.A.: Interface between globalization and technology. Asian J. Manag. Sci. **7**, 1–6 (2018)

Chapter 24
Empowering Vocational Education Ecosystem Through 5G+: Framework Design and Practical Approaches

Dawei Tao, Lan Bai, and Shih-Pang Tseng

Abstract The practical training component plays a crucial role in modern vocational education by bridging the gap between theory and practice, connecting schools with enterprises, and aligning education with employment. Establishing an efficient and sustainable training ecosystem is therefore of paramount importance. This chapter focused around the core of 5G+ technology and explores its comprehensive empowering role within the vocational education training ecosystem. The chapter first constructs a framework for the training ecosystem based on six elements: digital infrastructure, data analytics, personalized teaching resources, learning subjects, assessment and feedback mechanisms, and institutional policies. Through an in-depth analysis of how 5G+ technology operates at various levels and stages, the chapter demonstrates its critical role in achieving system adaptability, flexibility, and sustainability. In comparison to traditional training models, this ecosystem not only strengthens collaboration and interaction among participants but also facilitates a comprehensive improvement in educational quality and efficiency. This research provides an innovative theoretical and practical framework for the field of vocational education training, contributing to the digital transformation of vocational education.

D. Tao · L. Bai
Changzhou College of Information Technology, Changzhou, China
e-mail: taodawei@czcit.edu.cn; bl@czcit.edu.cn

S.-P. Tseng (✉)
Changzhou College of Information Technology, Changzhou, China

School of Information Science and Technology, Sanda University, Shanghai, China
e-mail: tsengshihpang@czcit.edu.cn

S.-P. Tseng et al. (eds.), *Advances in Intelligent Information Hiding and Multimedia Signal Processing, Volume 2*, Smart Innovation, Systems and Technologies 416, https://doi.org/10.1007/978-981-97-8760-9_24

24.1 Introduction

In the past decades, the dramatic development of information technology (IT) has changed the education environment. On the other hand, the globalization and Industry 4.0 have changed the needs of the high vocational education [5]. In recent years, the high vocational education institutes actively respond to the trends of digitalization, networking, and intelligence, driven by the Industry 4.0 [1]. The integration of "information technology" and "vocational education" has become the key tasks of the high vocational education institutes. Through the continual exploration and practice, an "IT-driven" education model with balanced teaching resources and personalized learning is gradually formed.

As a key link that can closely connect with industry needs improves students' job adaptability and organically combines knowledge and skills, practical training plays an irreplaceable role in cultivating the professionalism and workplace competency of vocational college students [3, 4]. The traditional vocational education training system faces two major challenges in the process of digital transformation: first, "uncertainty of teaching effectiveness," that is, due to the lack of efficient evaluation and feedback mechanisms, teaching effects are often difficult to quantify and optimize; second, "The low effectiveness of resource utilization," due to the time and space limitations of teaching resources and practical training environment, the breadth and depth of practical training teaching are restricted. The vocational education training system urgently needs to rely on 5G's open network and massive data, as well as advanced technologies such as artificial intelligence, virtual reality, and augmented reality to build a dynamically adaptable, highly interconnected, and virtuous cycle training ecosystem [2, 6].

Based on 5G+ technology empowerment and innovation needs, we try to build a vocational education training ecosystem. Through the construction of digital training application scenarios, we will promote the integration of industry and enterprise resources into the practical training teaching system and promote high-quality practical training education resources to balance regional differences, develop scenario-based, experiential learning, and intelligent education management evaluation, realize the transformation from skill-oriented to equal emphasis on comprehensive quality and innovation ability, and move from single vocational skills training to comprehensive educational goals.

24.2 Value Implications of "5G+" for the Construction of Vocational Education Practical Training Ecosystem

5G+ technology serves not only as a medium but also as a strategic tool leading the vocational education practical training ecosystem forward. By constructing a multidimensional, high-quality, collaborative, scalable, and self-correcting evaluation system, 5G+ technology becomes a key force driving the inherent virtuous cycle within the practical training ecosystem. Its value lies not only in the advanced nature of the technology itself but also in how this comprehensive technological framework enables a holistic, adaptive transformation of vocational education from teaching and practice to assessment.

24.2.1 Content Transformation: Multidimensional, High-Quality Practical Training Experience

The introduction of 5G+ technology injects new energy into the content of vocational education practical training. Traditional modes of practical training are constrained by physical environments, while 5G+ technology, with its low-latency and high-speed data transmission capabilities, constructs training scenarios using advanced technologies such as virtual reality (VR) and augmented reality (AR). These scenarios closely resemble real work environments, making them more suitable for simulating complex and high-risk practical training tasks. This shift enhances students' experiences from a singular dimension to a multidimensional one. High-quality practical training experiences manifest in students' in-depth understanding of the training content. While traditional training often focuses on operational aspects, the real-time data analysis and feedback mechanisms of 5G+ technology make it possible for students to delve into the theoretical aspects during the training process. Teachers can address individualized needs precisely based on real-time data, creating a dynamically adaptive teaching environment. This comprehensive content transformation will be a significant driver in shifting vocational education from skills training to comprehensive qualities and innovative capabilities.

24.2.2 Form Transformation: Collaborative, Scalable Training Models

The introduction of 5G+ technology elevates the compatibility of vocational education practical training forms. 5G+ technology supports highly interactive multiuser collaborative environments, and diverse teaching resources and training platforms across different regions can be more closely connected. Efficient and seamless collaboration is achieved among teachers, students, and training bases in different geographical locations. With its high-bandwidth and low-latency characteristics, 5G+ allows rapid completion of complex experiment simulations and large-scale data analysis, facilitating flexible integration of external resources such as industry experts, experimental equipment, and market demands. This profound transformation broadens the bridge between teaching and practice, theory, and application, providing a more flexible and sustainable training model that robustly supports the sustainable development of the vocational education practical training ecosystem.

24.2.3 Evaluation Transformation: Self-Correction and Strategically Adaptive Training Ecosystem

5G+ technology not only leads the transformation of training content and forms but also propels a new type of evaluation mechanism, self-correction with ecological features and strategic adaptability. The integration of 5G and big data analysis embeds the training system with the capability of self-correction. The system continuously monitors its operational status, such as the effectiveness of resource allocation and applicability of teaching methods, and automatically adjusts and optimizes through algorithms to ensure the overall efficiency and sustainable development of the entire training ecosystem. Additionally, the evaluation system possesses a strategic characteristic adaptable to external influences. Through real-time data analysis and feedback, the system not only makes personalized adjustments to individual student performance but also dynamically adapts strategies based on external environmental factors, such as changes in market demands and the emergence of new technologies. These two aspects of the evaluation mechanism combine to form an evaluation system with ecological features, enhancing the adaptability of the training ecosystem and endowing it with self-repair and updating capabilities, a true reflection of ecological characteristics.

24.3 Framework Design of "5G+" for the Construction of Vocational Education Practical Training Ecosystem

In exploring the framework of the vocational education practical training ecosystem, it is necessary to first clarify the model of traditional vocational education practical training and conduct a comparative study between the two. The traditional vocational education practical training system consists of six elements, including learning subjects and administrative and managerial support, showing a hierarchical and linear logical relationship among them (see Fig. 24.1). Under specific resources and environmental conditions, these elements can operate in an orderly and efficient manner, demonstrating stability and predictability. However, this top-down, highly centralized, and hierarchical model also reveals certain rigidity and limitations when dealing with rapidly changing external environments, technological advancements, or educational demands.

The framework for constructing the vocational education practical training ecosystem encompasses multiple interconnected aspects, including digital infrastructure, data analysis with dynamic adaptive algorithms, personalized and intelli-

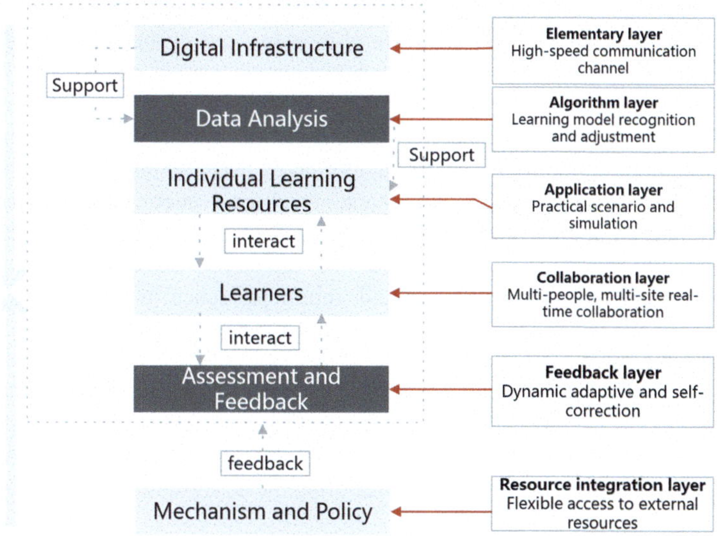

Fig. 24.1 Traditional vocational education practical framework

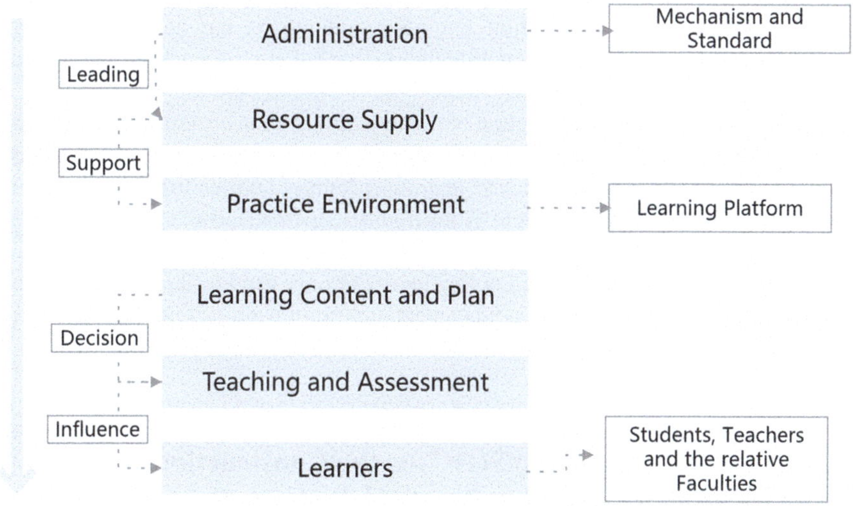

Fig. 24.2 "5G+" for the construction of vocational education practical training ecosystem

gent teaching resources, learning subjects, and evaluation and feedback mechanisms (see Fig. 24.2). In this complex and sophisticated system, "5G+" technology, by expanding the application scenarios of the practical training ecosystem in low latency, network slicing, and massive machine-type communications (mMTC), empowers each aspect like "nutrients." More importantly, it provides a new dimension to the logical connections between aspects, presenting the entire system with ecological diversity and sustainability. “5G+” plays a crucial role in the layers of data analysis with dynamic adaptive algorithms and evaluation and feedback mechanisms.

The data analysis with dynamic adaptive algorithms layer serves as the “nervous system” of the ecosystem, where “5G+” creates a high-speed, low-latency information flow channel in the digital infrastructure layer, accelerating the computational speed of the “nervous system.” It enables real-time interaction with learning subjects (students, teachers, industry experts), achieving personalized and intelligent allocation of teaching resources. This real-time interaction and dynamic adjustment are manifestations of “self-regulation” and “interdependence” within the ecosystem.

The evaluation and feedback mechanisms layer can be viewed as the ecosystem’s “self-regulator.” The data analysis capabilities of “5G+” technology enable this layer to acquire and analyze the performance and needs of learning subjects in real time, gradually self-correcting and optimizing. This not only enhances the quality of education but also provides more precise and real-time decision-making basis for the institutional and policy layers, demonstrating the ecosystem’s “adaptability” and “continuous evolution.”

By excellently performing at critical levels, “5G+” technology substantively promotes the interaction and adaptability of elements within the vocational education practical training ecosystem, truly endowing the system with “ecological” attributes. This logical connection not only strengthens the collaboration and interdependence among elements within the system but also brings higher resilience and sustainability to the entire ecosystem.

There is a bidirectional interaction between learning subjects (students, teachers, industry experts) and personalized and intelligent teaching resources. Teachers and industry experts can access students’ learning progress and feedback in real time, while students can directly benefit from highly personalized teaching resources. All elements within the ecosystem operate within the larger context of institutions and policies. The institutional and policy layer not only receives feedback from the entire system for corresponding adjustments and optimizations but also efficiently connects with other levels through advanced technologies like “5G+.”

24.4 Implementation of “5G+” for the Construction of Vocational Education Practical Training Ecosystem

The continuous advancement of “5G+” technology plays a proactive role in the reform and improvement of the vocational education practical training ecosystem. It collaborates with institutional and mechanism innovations, educational and teaching reforms, and the development of faculty teams to strengthen information sharing and communication, achieving the goal of “comprehensive development of vocational education, improvement of talent training quality, and adaptation to economic and social development.” This collaborative effect strengthens the interdependence and adaptability between various links and levels within the practical training ecosystem, laying a solid foundation for achieving the aforementioned goals. Through the comprehensive application and empowerment of "5G+" technology, the vocational education practical training ecosystem can efficiently deliver teaching, achieve multiple goals such as "resource optimization, quality assurance, and sustainable development," and ensure a high-level transformation of the system. To ensure the successful integration of "5G+" technology for the advanced development of the vocational education practical training ecosystem, it is essential to clarify and detail the design of practical pathways, ensuring smooth connections between various stages and levels. Specifically, from the perspectives of infrastructure, resource sharing, teaching models, teaching strategies, collaborative alliances, and assessment and feedback mechanisms, the implementation of the practical pathway should be gradually advanced, forming an organic and collaborative vocational education practical training ecosystem.

24.4.1 Infrastructure—Integration of "5G+" into Vocational Education Ecosystem

The establishment of infrastructure is the primary step in the practical pathway and serves as the cornerstone for constructing the "5G+" vocational education practical training ecosystem. It reflects the important manifestation of the fourth-generation information technology revolution's "cross-border" and "integration" characteristics in the field of vocational education. The infrastructure based on "5G" technology includes stable high-speed network connections and the establishment of smart cities and smart campuses. These infrastructures facilitate information exchange, resource sharing, data exchange, and system linkage, thereby improving the overall utilization efficiency of vocational education resources and ensuring the reliability and stability of the vocational education practical training ecosystem. The construction of infrastructure should be guided by the needs of the "5G+" vocational education practical training ecosystem, integrating and optimizing the integration of hardware facilities and software applications to enhance the overall efficiency of the system. Due to the high-bandwidth and low-latency characteristics of the 5G network, upgrades are required for existing network devices such as routers and switches. Additionally, enhancements in cloud computing facilities' computing and storage capabilities are necessary to meet higher network transmission speeds and larger data processing volumes. Consideration should also be given to the incorporation of virtual reality (VR), augmented reality (AR), Internet of Things (IoT) devices, and applications to ultimately serve the vocational education practical training ecosystem.

24.4.2 Resource Sharing—Optimization of Resources in the "5G+" Vocational Education Practical Training Ecosystem

Resource building and sharing are core components of the "5G+" vocational education practical training ecosystem. "5G+" technology optimizes various teaching resources through network technology and emphasizes dynamic optimization and personalized configuration of resources, including but not limited to digital textbooks, virtual simulation experiment simulation software, VR/AR content, etc. In the vocational education practical training ecosystem, "5G+" achieves efficient resource allocation, dynamic scheduling, and real-time sharing, improving the utilization efficiency and teaching effectiveness of educational resources. Simultaneously, resource building and sharing under the "5G+" environment also focus on the introduction of advanced technologies such as artificial intelligence to provide intelligent resource recommendations and personalized learning paths, enhancing the learning experience and effectiveness of vocational education.

The process of achieving resource building and sharing in the "5G+" vocational education practical training ecosystem is highly strategic and operational. First, precise analysis of the diverse demands of vocational education is necessary to determine the required resource types and content. Second, leveraging the high-speed transmission and big data processing capabilities of 5G technology, existing educational resources should be integrated and optimized to enhance the speed and efficiency of resource storage and distribution, meeting the demand for users to access resources quickly and efficiently. Effective resource building and sharing are not a one-time task; a continuous resource update and maintenance mechanism should be established to ensure the quality and currency of resources. This includes regularly collecting and analyzing user feedback, understanding the effectiveness of resource utilization, and making timely improvements or replacements for resources that need optimization.

24.4.3 Teaching Models—Innovation of Teaching Models in the "5G+" Vocational Education Practical Training Ecosystem

Innovating teaching models is critical to the "5G+" vocational education practical training ecosystem. As proposed in the "2030 Future Learning Guidelines," the construction of a community learning model that breaks the boundaries of school walls will have a fundamental impact on vocational education. The transformation of vocational education models to scale, standardization, and intelligence, breaking the time and space constraints of traditional teaching, enables intelligent teaching modes, online collaborative learning, and community learning, addressing opportunities and challenges brought by large-scale industrial upgrades.

The community learning model built on "5G+" technology serves as an effective complement to vocational school teaching. First, vocational schools can collaboratively create an online interactive learning environment with communities using "5G+" technology. This environment can provide customized learning paths for students and provide feedback on students' learning to educators, facilitating educators in adjusting teaching strategies. Members of the community can directly interact with students in this environment, providing students with more practical learning and practice opportunities. Second, "5G+" technology helps achieve the sharing of educational resources. Vocational schools and communities share educational resources in this ecosystem, such as opening vocational school course resources to the community, and the community opens up practical opportunities to vocational schools. This way of sharing educational resources not only improves the utilization rate of educational resources but also brings vocational education learning and practice closer to real environments.

24.4.4 Teaching Strategies—Designing Strategies in the "5G+" Vocational Education Practical Training Ecosystem

Teaching strategies play a decisive role in the "5G+" vocational education practical training ecosystem. With the introduction of "5G+" technology, teaching strategies have evolved to be more intelligent and personalized, further optimizing the teaching process. In this process, customization of personalized learning paths for a broader range of students, real-time feedback on teaching progress and effectiveness, and the autonomous development of intelligent learning tools become crucial. Leveraging the capabilities of big data and artificial intelligence, "5G+" technology provides students with personalized learning paths, real-time feedback on teaching progress and effectiveness. This implies recommending the most suitable learning resources and tasks based on students' learning habits, abilities, and interests. Additionally, it provides teachers with intelligent feedback to timely adjust teaching strategies. Furthermore, teaching support platform tools based on "5G+" technology are more technologically compatible. Teachers can autonomously develop intelligent learning tools tailored to specific vocational education practical training environments, simulating skill learning and practice in real-world job environments. This makes the teaching process more vivid, practical, and enhances teaching quality and effectiveness.

24.4.5 Collaborative Alliances—Cross-Border Collaboration in the "5G+" Vocational Education Practical Training Ecosystem

Collaborative alliances form the foundation of the "5G+" vocational education practical training ecosystem. "5G+" technology, in the vocational education practical training ecosystem, relies on cross-campus, cross-industry, and cross-disciplinary collaborative alliances as key building elements, following the guidance of cross-domain collaboration theory. This theory emphasizes shared collaboration of resources and knowledge to enhance resource efficiency, problem-solving capabilities, and stimulate innovation, aligning with the core goals of the practical training ecosystem. Various stakeholders, supported by the stable high-speed communication network of "5G+" technology, can exchange and share information in real time, facilitating collaborative communication, building understanding and trust, resolving conflicts, and ensuring smooth collaboration processes, thereby advancing high-quality and fair vocational education.

It is crucial to ensure data consistency and interoperability by constructing unified data formats and exchange protocols during the collaborative alliance building process. First, the responsibilities and rights of each participant in the alliance should be clearly defined and explicitly written into the collaboration agreement. This not only helps maintain the stability of the alliance but also aids in finding a basis for problem resolution in case of disagreements. Second, the formulation of unified data formats and exchange protocols is essential. Specifically, unified data models, data element definitions, and data exchange interface standards should be formulated based on the data exchange requirements, ensuring data consistency and interoperability. For data exchange interface standards, existing open standards such as RESTful API or GraphQL can be adopted to simplify the data exchange process and ensure data consistency. Finally, data security and privacy protection are critical factors that must be considered. When formulating data exchange protocols, the permissions of each party regarding data use should be clearly defined, and relevant data security and privacy protection policies should be established to prevent data leakage or misuse, ensuring the rights of all participants.

24.4.6 Effect Evaluation—Data Feedback in the "5G+" Vocational Education Practical Training Ecosystem

Effect evaluation and feedback are crucial elements in the "5G+" vocational education practical training ecosystem. The "5G+" technology-based vocational education practical training ecosystem demonstrates innovative practical patterns in setting evaluation indicators, developing evaluation tools and systems, and constructing teaching feedback and improvement mechanisms. First, the introduction of technologies such as artificial intelligence and machine learning mines learning patterns and trends from students' learning behaviors and feedback, further establishing a comprehensive and personalized evaluation indicator system. These indicators may include students' learning progress, learning effectiveness, participation, and satisfaction, aiding in a more precise understanding and guidance of students' learning. Second, "5G+" technology is utilized to develop real-time dynamic evaluation tools and systems, monitoring students' learning behaviors in real time, providing quick feedback on learning progress and effectiveness, supporting micro-adjustments to teaching quality with data. Lastly, the establishment of a real-time, transparent, and multidirectional teaching feedback and improvement mechanism based on "5G+" technology involves regularly consulting and analyzing learning data, adjusting teaching strategies promptly based on feedback information. Simultaneously, students are encouraged to actively provide feedback on their learning experiences within this mechanism, achieving continuous optimization of teaching goals.

24.5 Conclusion

Building a vocational education and training ecosystem based on 5G+ technology involves three key levels: the foundational infrastructure, the intermediate applications, and the top-level feedback. The foundational infrastructure provides essential technological and resource support, ensuring the stable operation of the vocational education ecosystem. The intermediate applications, built upon the foundational infrastructure, involve practical applications that enhance the education process through innovative teaching modes and strategies, making education more personalized and intelligent. The top-level feedback serves as a basis for system optimization and improvement, continuously enhancing teaching quality and effectiveness through evaluation and feedback mechanisms.

Therefore, the construction of a vocational education and training ecosystem requires simultaneous attention to building the foundational infrastructure, innovating intermediate applications, and establishing top-level feedback. Only when these three levels are coordinated and mutually supportive, can a complete and efficient vocational education support system be formed, providing students with high-quality education and training services. In the context of a 5G+ technology-enabled vocational education ecosystem, advancing the development of foundational infrastructure, intermediate applications, and top-level feedback is essential for achieving sustainable development and continuous innovation in vocational education.

Acknowledgments This work was partially supported by the Changzhou College of Information Technology under the contract "2019KYQD03." And this work was also partially supported by Sanda University under the contract "2021ZD05."

References

1. Ahmed, I., Jeon, G., Piccialli, F.: From artificial intelligence to explainable artificial intelligence in Industry 4.0: a survey on what, how, and where. IEEE Trans. Industr. Inform. **18**, 5031–5042 (2022). https://api.semanticscholar.org/CorpusID:246366040
2. Dangi, R., Lalwani, P., Choudhary, G., You, I., Pau, G.: Study and investigation on 5G technology: a systematic review. Sensors (Basel, Switzerland) **22**, 26 (2021). https://api.semanticscholar.org/CorpusID:245465682
3. Huai-ya, J.: Study and practice of practical training mode in the course of architectural mechanics in high vocational education. Career Horizon (2013). https://api.semanticscholar.org/CorpusID:115061813
4. Nyen, T., Tønder, A.H.: Development of vocational skills through integration of practical training periods in school based vocational education in Norway (2018). https://api.semanticscholar.org/CorpusID:158816255
5. Xiong, H., man Chang, K.: The impact of vocational education on the high-quality development of local economy in the new era. Adv. Vocat. Techn. Educ. **4**(4), 63–69 (2022). https://api.semanticscholar.org/CorpusID:254364341
6. Zhongjun, T., Shah, S.K., Ahmad, M., Mustafa, S.: Modeling consumer's switching intentions regarding 5G technology in China. Int. J. Innov. Technol. Manag. **19**(04), 2250011 (2022). https://api.semanticscholar.org/CorpusID:248714241

The manufacturer's authorised representative in the EU is Springer Nature Customer Service Centre GmbH, Europaplatz 3, 69115 Heidelberg, Germany. If you have any concerns regarding our products, please contact ProductSafety@springernature.com

Printed and bound by CPI Group (UK) Ltd, Croydon, CR0 4YY
01/10/2026
02225958-0004